the
green
methods

M A N U A L

The original bio-control primer

by Michael S. Cherim

Edition IV for 1998

Many thanks to those who helped
make this publication possible by providing
critical materials, technical information,
your support and understanding.

This book is dedicated to my family, friends and customers.

The Green Methods™ Manual, The original bio-control primer

Copyright © 1998 by Michael S. Cherim

All rights reserved. No part of this publication may be reproduced or transmitted in any form or by any means, electronic or mechanical, without the written permission of the author, except for short passages used in critical reviews.

ISBN: 0-9648682-0-2
Price: $9.95 U.S. - Stock Code: EGMM4
Library of Congress Catalog Card Number: 96-94578
First Printing - 1998

For queries and/or comments, the author may be contacted through to the office of the publisher (send correspondence to the address below). This technical manual was published and is distributed by:

The Green Spot, Ltd.
Publishing Division
93 Priest Road
Nottingham, NH 03290-6204

TEL: 603/942-8925
FAX: 603/942-8932

This publication was bound and printed in the United States of America on recycled paper. Every reasonable effort was made to minimize impact to the environment during the entire publication process.

A portion of this publication's title: "Green Methods" is, and has been since 1993, a copyrighted trademark of the publisher. It has been used in the title and other parts of this publication with permission.

Disclaimer: Every effort was made to provide concise and accurate details of biological pest control. However, due to the nature (nature being the operative word) of biological pest control, many variables exist. It is therefore recommended that practitioners of bio-control do follow the procedures and protocols explained in this text only with a certain amount of reservation and caution. Only small undertakings are recommended for novices, until, through experimentation, the techniques best suited for that person, system or operation are revealed and a certain amount of confidence is realized. Neither the author nor the publisher will assume responsibility for injuries, damages or losses incurred as a result, direct or consequential, of the information contained in this publication.

The author has received no special compensation for his work on this manual. He is, however, a paid employee of the publisher. Compensation to the publisher is made not through the sales of this book, but through the sales of the products described herein. Your patronage is greatly appreciated.

Table of Contents

Other Products Used

The Business End

forward

Dear Reader,

To those of you who are new to The Green Spot, we welcome you! And to those who already know us, we're back and catalog-free (unless you count this tech manual with pricing).

Some of you — hopefully not too many — will remember our slightly premature revelation to an up and coming tech manual. Well it never got started and the project was abandoned. It sort of was, anyway, we had one already: our Green Methods catalog. With a few modifications... well, does the expression, "kill two birds with one stone," mean anything? It sure did to us!

Among the text in this manual you'll find sections dedicated to the sales of the products discussed. Additionally, you'll find a section which discusses the workings of purchasing product, shipping, etc.

We certainly hope everyone likes this format. Hopefully it is not revealing too much sloth. If it is... hey, what can we say? We're saving trees.

One last thing before we delve into this publication, a troubling thing at best, what to do about the expense of this book? Some of you will receive this manual at no cost to you. If you are a regular paying customer: enjoy the book, it's on us. However, if you're new to us (especially if you're not an industry business) we'd sure appreciate a check for $9.95 to cover our costs.

If you paid for this manual outright, as a response to an ad or something, thank you very much, your contribution is much appreciated. If you've received this book and lack interest in bio-control and IPM (integrated pest management), maybe you could do us a favor: give it to a local grower, garden center, etc., and let us know so that we make take your name off our list or, perhaps, put the account into the name of the new recipient.

Running an honest, forthright business requires a set of expenditures not normally considered: such as a technical manual (we could simply sell our products in a 32 pager and save ourselves a bundle). So every time we thank you for your consideration and appreciation, we really do mean it. Hope you enjoy the book!

Mike Cherim (the bug guy)

Green

Methods

101

our
introductory
section

photo, opposite, by M. Cherim

introduction - part one

There is no doubt about it, bio-control works. It's not a cake-walk. It's not usually the least expensive form of pest control (unless you consider all the factors — especially labor — then you're in the ball park with chemical means). And environmentally, you're light years ahead of the game.

In order for bio-control to be effective to a degree equivalent to or greater than that of chemicals, you must first learn some of bio-control's peculiarities — as you did before you applied your first spray. It is true: learning by doing is a lesson well remembered. However, with bio-control, preparation is really fundamental. In more ways than one...

The first thing you should know, to make the whole experience as painless as possible, is to accept that multiple practices should be studied simultaneously. The reason: bio-control can, but often-times shouldn't be, used as a sole means of pest control. A different approach is often needed. One which utilizes and integrates several pest management resources. This is referred to as Integrated Pest Management, or IPM, for short. Pest management resources available to IPM and bio-control practitioners include the following (in no particular order):

The bio-control agents. Many organisms, often called "beneficials" are available to growers, interiorscapers, farmers and others. There is a pretty fair assortment right in this manual. Beneficials are comprised of two basic groups which are divided by their methodology: some are predatory; some are parasitic (and some do do a little [host-feeding] of both).

Predacious beneficials can include insects such as beetles, wasps and true bugs. Certain mites and other critters also share this characteristic.

Parasitic beneficials, on the other hand, consist chiefly of mini-wasps (parasitoids) and nematodes. Many commercially available organisms, belonging to both groups, will be discussed in greater detail throughout this manual.

Pesticides. The beneficials commercially produced today have certain inherent and, sometimes, engineered tolerances to certain pesticides. (By pesticides, we mean synthetic, natural or botanically derived insecticides, acaricides or miticides, nematicides, etc.) Tolerances to these substances, or resistance as it is known, is a trait common to most organisms, to a lesser or greater degree, by species. It is especially prevelent among pests due to continued exposure.

There are a reported 447 species of pests which have an unusually high tolerance of a great many —cides. These pests have been called "super bugs" and their resistance is due to being repeatedly subjected to a single class of chemical — one class at a time. And often it is due, in conjunction, to ineffective and improperly applied substances.

As a general rule, predators and parasitoids do not achieve the level of tolerance common to so many pests; they're not subjected to the substances as often and avoid treated areas. Therefore caution should be exercised if chemicals are used as a portion of an IPM program. If these substances can be left out of an IPM program, that program has a far greater chance of sustainability. Moreover, as an incentive to leave the sprays in the storage closet, growers have been reporting significant —cide effectiveness when generally removed from the program, then reinstated temporarily — and rarely — to address a major pest flare-up when and if one occurs.

If —cides are going to be a part of your pest management program, it is advisable to address the needs of your beneficials first. Certain substances are compatible with certain bio-control agents; some substances more compatible than others; some have a shorter residual effect than others and, despite their toxicity, have a place in an IPM program if they are to be used judiciously and sparingly. And that's a good rule to follow when employing natural enemies. For more on biorational substances, see fig. 1 (next page). Also, check out our selection of these items As far as pests becoming resistant to beneficials? It can't happen. Worst case scenario: the pests evolve, slowly developing stronger, more predator-proof armor — or something like that, anyway.

Physical means. Handpicking is physical. It is 100% effective against the pests caught in the pinch. Even on a grand scale, if the timing is right and the scouting is first-rate, handpicking can be an extremely effective tool. Our recent experiences confirm this. If the timing isn't right, though, it might be less expensive in some cases to discard the crop than it would be to pay the pest-picker's price.

Physical means mean much more than just handpicking, though. You can also incorporate the use of a physical control devices and products: insect screens; sticky, slippery, charged or abrasive barriers; row covers, scent barriers and more. We are limited only by our imagination. Some of these products are covered in this manual in the Physical Controls section.

Continued on page 7...

*There are many high-quality organic fertilzer products and expert
advice about soil nutrition and soil science available from:
North Country Organics
R.R. 1, Box # 2232, Bradford, Vermont 05033
Ask for their complete and detailed literature pack, call: 802/222-4277

biorational substances chart

FIG. 1

This is a quick reference chart of our organisms and the substances which are compatible with them. **Compatibility** means that the mortality rate of said organism is <25% after 24 hours of exposure. It is still best to exercise **Caution** and **Moderation** when applying these substances around the organisms, as some losses will occur. Moreover, the product's inert ingredient(s) may prove harmful. For info pertaining to effective residual periods for these and other products, please call.

BIO-CONTROL AGENT	COMPATIBLE PRODUCT CODES	KEY TO PRODUCT CODES
Aphidius colemani...............	*1, *1a, 2, 4, *6, 7, 8, 9, *10, *11, 12, 13, 14, 15, 16, *17, 18, 19, 21, 24, 27, 28, 31, *31a, 37, 40, 41, *41a, *43, *43a, 44, 45, *45a, *47, *47a, *52, 54,*55, 56, 57, *58, *59, *63, 66, *67	1 = Aaterra 1a = Actinovate 2 = Agrimycin, Agri-Strep 2a = Aliette
Aphidius matricariae...........	*1, *1a, 2, 4, 6, 7, 8, 9, *10, *11, 12, 13, 14, 15, 16, *17, 18, 19, 21, 24, 27, 28, 31, *31a, 37, 40, 41, *41a, *43, *43a, 44, 45, *45a, *47, *47a, *52, 54,*55, 56, 57, *58, *59, *63, 66, 67	3 = All organophosphates 3a = Ambush 4 = Apollo
Aphidoletes aphidimyza.......	*1a, 2, 4, *7, 8, 9, 12, 13, 14, *15, 16, *17, 18, 19, 21, 24, 26, 27, 28, *31a, 35, 37, 41, *41a, *43, *43a, *45a, 47, 47a, 56, 66	5 = Applaud 6 = Avid 7 = Azatin XL (neem)
Chrysoperla spp.................	*1a, *2, *4, *7, *8, *12, *13, *15, *21, *27, *28, *31a, *33, *35 *43, *43a, *45a, *56, *66	8 = B-Nine 9 = Basamid 10 = Bavastin
Deraeocoris brevis..............	*1a, *2, *4, *7, *8, *12, *13, *15, *21, *27, *28, *31a, *33, *35 *41a, *43, *43a, *45a, *47a, *56, *66	11 = Baycor 11a = Bayleton 12 = Benlate
Hippodamia convergens.....	*1a, *2, *4, 7, *8, *13, *15, *19, *21, *27, *31a, *33, *35, *37, *41, *41a, *43, *43a, *45a, *48, *56, *66	13 = Botran 14 = Bravo 15 = Bt-k, B, SD, H-14, Dipel
Trichogramma spp..............	*1a, *2, *4, *7, *8, *12, *13, *15, *19, *21, *27, *28, *31a, *33, *37, *41a, *43, *45a, *47a, *56, *66	16 = Captan 16a = Carbaryl 90DF 17 = Copper Bordeaux mix
Parasitic nematodes...........	*All compatible except for nematicidal products	18 = Copper, fixed 19 = Cycocel
Dacnusa siberica...............	*1a, *2, *4, *7, *8, *12, *13, *15, *19, *21, *27, *28, *31a, *33 *35, *37, *41a, *43, *43a, *45a, *47a, *56, *66	20 = Cyprex 21 = Daconil 2787 22 = Diazinon FS
Diglyphus isaea.................	*1a, *2, *4, *7, *8, *12, *13, *15, *19, *21, *27, *28, *31a, *33 *35, *37, *41a, *43, *43a, *45a, *47a, *56, *66	23 = Dikar 24 = Dimilin 25 = Dithane
Aphytis melinus.................	*1a, *2, *4, *7, *8, *12, *13, *15, *19, *21, *27, *28, *31a, *33, *37, *41a, *43, *43a, *45a, *47a, *56, *66	26 = Easout 27 = Enstar 28 = Exotherm termil
Cryptolaemus montrouzieri..	*1a, *2, *4, 7, 8, *12, *13, *15, 19, 21, 27, 28, *31a, *33, *35, *37, 41, *41a, *43, *43a, *45a, 48, 56, 66	29 = Ferbam 30 = Fungaflor 31 = Funginex
Harmonia axyridis..............	*1a, *2, *4, *7, *8, *12, *13, *15, *19, *21, *27, *28, *31a, *33, *35, *37, *41a, *43, *43a, *45a, *56, *66	31a = Garlic Barrier 32 = Guthion 33 = Imidan
Metaphycus helvolus..........	*1a, *2, *4, *7, *8, *12, *13, *15, *19, *21, *27, *28, *31a, *33, *35, *37, *41a, *43, *43a, *45a, *47a, *56, *66	34 = Insecticidal soap 35 = Karathane 36 = Kelthane
Rhyzobius lophanthae.........	*1a, *2, *4, *7, *8, *12, *13, *15, *19, *21, *27, *28, *31a, *33, *35, *37, *41, *41a, *43, *43a, *45a, *48, *56, *66	37 = Kumulus 38 = Lorsban (1/2 rate), Dursban 39 = Manzate
Neoseiulus fallacis............	*1a, *2, 3, *3a, *4, 6, 7, *8, 11, 12, *13, *15, 16, *17, *18, *19, 20, *21, 23, 25, *27, *28, 29, *31a, 32, 33, 36, *37, 38, *41a, *43, *43a, *45a, 46, 47a, 48, 50, 51, 55, 57, 58, 61, 62, 63, 65, 66	40 = Manzate 200, Daronil 200 41 = Meltatox, Milban 41a = Milky spore dis., B. popilliae
Neoseiulus fallacis PR........	Same as above plus key code 53	42 = Morestan, Joust 43 = Mycostop
Phytoseiulus persimilis.......	*1a, 2, 2a, 4, 5, 6, 7, 8, 9, 10a, 11, 13, 14, 15, 16, *17, 18, 19, 21, 22, 24, 25, 27, 28, 30, *31a, 33, 35, 37, 39, 41, *41a, *43, *43a, 44, 45, *45a, 47a, 48, *49, 52, 56, *58, 59, 60, 61, 64, 66, 67	43a = B. bassiana 44 = Nimrod 45 = Nissorun
Stethorus punctillum..........	*1a, *2, *4, *7, *8, *12, *13, *15, *19, *21, *27, *28, *31a, *33, *35, *37, *41a, *43, *43a, *45a, *56, *66	45a = Nosema locustae 46 = Omite, Ornamite 47 = Pentac
Hypoaspis miles................	*1a, *2, 4, *5, 7, *8, *9, *13, *14, *15, *17, *18, *19, *21, *24, *27, *28, *31a, *33, *41, *41a, *43, *43a, *45a, 47a, *48, *56, *66	47a = Phyton 27 (copper) 48 = Pirimor 49 = Plant-Fume (nicotine)
Iphiseius degenerans.........	*1a, 2, 4, 7, 8, 9, 13, 14, 15, *17, 18, 19, 21, 24, 25, 27, 28, 31, *31a, *33, 35, 37, 39, 41, *41a, *43, *43a, *45a, 47a, 48, 56, *58, 66	50 = Plictran 51 = Polyram 52 = Previcur-N
Neoseiulus cucumeris.........	*1a, *2, 4, 5, 7, *8, *9, 10, 11, 12, 13, *14, 15, *17, 18, 19, 21, 22, 24, *25, 27, *28, 30, 31, *31a, *33, *34, 35, *37, *39, *41, *41a, 43, *43a, 44, 45, *45a, 48, *49, 55, 56, 57, *58, 60, *63, 64, 66, 67	53 = Pyrethrum, Decathlon 54 = Ridomil 55 = Ronilan
Orius insidiosus.................	*1a, *2, *4, 7, *8, *12, *13, *15, 19, *21, *27, *28, *33, *35 *41a, *43, *43a, *45a, *47a, *56, *66	56 = Rovral, Chipco 26019 57 = Rubigan 57a = Sevin XCR Plus, Gel, 50w & 4F
Delphastus pusillus...........	*1a, *2, *4, *7, *8, *12, *13, *15, *19, *21, *27, *28, *31a, *33, *35, *37, *41a, *43, 43a, *45a, *56, *66	58 = Sulfur 59 = Sumisclex 60 = Supareen-M
Encarsia formosa..............	*1, *1a, 2, 4, 7, 8, 9, *10, *11, 12, 13, 14, 15, 16, *17, 18, 19, 21, 24, 27, 28, 31, *31a, 37, 40, 41, *41a, *43, *43a, 44, 45, *45a, *47, 47a, *52, 54,*55, 56, 57, *58, *59, 63, 66, 67	61 = Talstar 61a = Tedion 62 = Thiodan, floor spray
Eretmocerus eremicus......	*Use key for E. formosa -- it should be accurate	63 = Thiram 64 = Topsin M
Bumblebees.....................	*1a, 4, 7, 12, *15, 34, *41a, *43, *43a, *45a, *47a, 48, 56, 58, 59, 67	65 = Trygard 66 = Vanguard
Fly parasites....................	*1a, *2, *4, *7, *8, *12, *13, *15, *19, *21, *27, *28, *31a, *33, *35, *37, *41a, *43, *45a, *56, *66	67 = Vendex, Torque plus 68 = Zineb
Tenodera aridifolia sinensis..	*1a, *2, *4, *7, *8, *12, *13, *15, *21, *27, *28, *31a, *33, *35 *41a, *43, *43a, *45a, *56, *66	

The information in the chart above was provided by Applied Bio-Nomics, Ltd., Sidney, B.C. Canada, Biobest Biological Systems, bvba, Westerlo, Belgium, Koppert B.V., Netherlands and Dan Mayer, Wash. State Univ. Information about the biorational substances for use with Amblyseius fallacis was compiled from slide-dip testing experiments conducted by S.E. Lienk, N.Y.S. Agr. Expt. Sta., Geneva, N.Y.

Plant culture. *Like that which is true of humans, the immune system being more at risk during times of illness — plants, too, are more susceptible to pests and diseases when they are under duress. Health and vigor are often enough to adequately protect a plant from its enemies. Fertilization materials and techniques, aeration, irrigation, light, humidity and temperature can all influence plant health. They all, therefore, influence pest control and disease management, too. We recommend the use organic fertilizers where possible and practical. *Organic fertilizers (see page 5) lend themselves to the health of a plant by enriching the soil or medium. Not by feeding the plant directly. Organic fertilizers, of which there are many such as composts, minerals, animal by-products, etc., are usually long-lasting, non-burning, and well-balanced with slow-released macro- and micronutrients. And of the most importance, perhaps, where applicable, they don't destroy the soil ecosystem whose members perform tasks essential to life on earth.*

Trapping. *This, too, is a physical control device, but warrants a paragraph of its own. Trapping is usually implemented for one of two reasons: to detect the presence of, and determine the identity of, pests; or to significantly reduce to the pest population by trapping great numbers of the pest or pests of concern.*

Using one 3" x 5" yellow sticky trap per every 250 sq. ft. of floor or bench space could be useful to monitor a crop for whiteflies, fungus gnats, thrips, leafminers, aphids and a host a other organisms, including many beneficial species. However, using 1 trap per sq. yd. might actually lend itself to the control of those organisms. Unfortunately, trap numbers that high would probably decimate a huge chunk of the available parasitoids and predators.

Another way to monitor-trap would be to use a trap that drew large numbers of a specific pests' gender. For example: using a pheromone or sex scent of a particular female pest, in theory, will lure males to a trap. Using a scent as a lure, however, unlike a visual trap such as colored sticky traps, may provide unrealistic numbers. And watch out for traps that lure both sexes. Males and females which are lured by means of multiple scent lures may not all, necessarily, fall victim to the trap, in which case they'll mate. And in which case, their offspring may wreak havoc on your plants.

*We discuss multi-sized yellow and blue visual sticky traps in this manual (in the Scouting Paraphernalia section). However, when it comes to using scent traps and the such, please refer to a *specialist in that field (see below).*

*For a large selection of traps, IPM accessories and solid info contact:
Great Lakes IPM
10220 Church Road NE, Vestaburg, Mich. 48891
Ask for their info-packed manual, call: 517/268-5693

Scouting. *Scouting should be considered a pest management resource. It is an expenditure of manpower. And, realistically, it is probably one of the most important contributing resources to a successful IPM program. It can also help establish the entire course for a well managed IPM program after a only a couple of years.*

Scouting, in order for it to be useful to an IPM program in this way, has to be carried out in an organized and focused fashion. Scouting should be performed at regular, timed intervals: once-a-week, on the same day, at the same time. No deviation should be acceptable. It is a good idea to make a map of the site. On the map should be a route with checkpoints along the way, say sticky traps or indicator plants. Observations of all anomalies should be recorded [on the back of the map]. A log of historic information should be maintained and graphed along the way; it can come in hand later on, as described below.

When you do scout and observe and record and recall, many trends can be detected. For instance: you notice that, historically, you get aphids every April, towards the end of the month. With this information an educated determination can be made: an aphid parasitoid, one which is useful as an inexpensive preventive, should probably be released in the crop starting at the beginning to the middle part of March. This practice will probably be less expensive than using the same mini-wasps to counter an in-progress aphid flare-up.

We discuss a few useful scouting items in this manual which you'll find you may need. For info about our consultation services, please call us.

Trap- and banker-crops. *These are heavily debated techniques of coping with pests. They are ones that we at The Green Spot are in favor of.*

A trap-crop consists of a number of close-proximity non-crop plants which are specially selected to lure, harbor and divert a majority population of a specific pest or pests. Moreover, because these plants lure and sustain the pests [hosts], they sustain the beneficials which are also lured to them. We've seen this employed several times successfully — and never unsuccessfully. Small flowering plants seem to work best: tansy, crimson clover, alyssum are some which work well. Umbelliferous flowers, such as dill, work

Ladybug visiting flowers of dill plant.

Photo by M. Cherim

well for ladybugs (see photo, facing page). We offer a special seed mix, "Bioblend," which contains a variety of plant seed suited to this purpose. In any case, trap-crop plants should be planted between the protected crop and the wildlands, or direction from which the pests are suspected of coming.

Banker-crops are slightly different. Banker-crops are used to sustain certain beneficials which do not normally propagate themselves effectively on the protected crop. One example: our predatory true bugs, Orius insidiosus and Deraeocoris brevis, don't perform too well on tomatoes (their progeny have difficulty negotiating the tomato plant's hairs). If a row of peppers are planted every fifth row, or interplanted, every fifth plant, O. insidiosus and D. brevis adults could perform well on the tomatoes while their young spend their early instars on the peppers. Pest levels in the greenhouse, on both crops, might be successfully maintained at low, inconsequential levels for the duration of the crop.

Education. Yes, this too should be considered a resource. Without the proper help, IPM can be a fail-likely practice. One should either explore-by-doing on a small scale, reserving the rest of the operation for conventional methods; or one should brush up on the subject before delving into its practical applications. We at The Green Spot try really hard to prepare first-time practitioners for some of the pitfalls and setbacks one might experience. And, through our company's literature, we make an attempt to educate our customers so that they may attain success (and we, of course, will have happy, loyal clients). Hopefully this manual, like our catalog, gets high marks in the education and enlightenment departments.

There are many, many ag and hort book titles available through this company's manual, and we feel that as a literature and information resource, they fit the bill:

AgAccess
P.O. Box 2008, Davis, Calif. 95617-2008
To request more information or receive their manual, call: 916/756-7177

This organization, of which The Green Spot is a member, concerns itself with virtually all significant aspects of North American and worldwide bio-control issues:

Association of Natural Bio-control Producers - A.N.B.P.
10202 Cowan Heights Drive, Santa Ana, Calif. 92505
To request more information about them or their newsletter, call: 714/544-8295

This organization, of which The Green Spot also is a member, concerns itself with virtually all new least-toxic pest management technologies:

Bio-Integral Resource Center - B.I.R.C.
P.O. Box 7414, Berkeley, Calif. 94707-7414
To request more information about them or their two publications, call: 510/524-2567

Aside from our literature, there are several decent books out there which will attempt to increase our knowledge of scouting, IPM, insect biology, disease management, arthropod identification, etc. (Arthropod identification is especially important to the scout. Identification guides and keys will help determine if the "bug" is good, bad or otherwise.) If you've no time to buy and read a bug book, The Green Spot does have an inexpensive and efficient arthropod identification service called Pro-Scope℠. This service has been helpful to many people. It has degenerated, though. We used to have the resources to tackle just about anything. Nowadays, though, we can only handle scale insect and whitefly identification projects (by scale insects we mean soft scales, mealybugs and, to a lesser degree, armored scales) unless it is something simple or if it is a request for a simple/ cursory identification.

Bio-control and IPM are becoming popular practice, the need for our Green Methods is growing into something magnificent. And with that popularity comes a demand for educational resources, outlets, discussions and sounding boards. We have several good titles that we offer through this manual, we also offer Pro-Scope, a free telephone-consultation service and other, more intense and not-quite-free, consultation services as well. But if that's not enough, try your local university's cooperative extension office or consider one of the resources shown at the bottom of the preceding page. ✄

If chemicals are the pest control answer...

Why didn't someone tell the bugs?

introduction - part two

Now that the mechanics have been covered — sort of, anyway — it is time to peer into the more obscure side of bio-control and IPM. It almost takes a kind of attitude to deal with the many facets of these products of agri-ingenuity.

People who practice bio-control often refer to their still-chemical-spraying industry-mates as having a "spray mentality." They say this not out of meanness, arrogance or spite, but it is brought-fourth by their knowledge of the difference between the preemptive mind-set of the bio-control practitioner and the reactive mind-set of the chemical applicator.

As with any discipline, bio-control and IPM adhere to myriad rules and protocols. Some of which require time to learn and understand. This is no different than understanding that certain chemicals cannot be tank-mixed or that horticultural oil used in strong sunlight may burn plant tissue. The Green Methods, as with so much in life, provides us with a challenge. Its mastery is an obstacle which should, really for our best interests, be met and overcome.

The Green Methods' facets consist of some of the following which are shown in their most basic form. Remember, when dealing with the Forces of Mother Nature, anything goes.

Prevention. The expression, an ounce of prevention is worth a pound of cure, is so relevant to the Green Methods. It is so often a necessity for success. In order to apply preventive measures, though, one must think about a problem before a problem exists. For some, this can be tremendously difficult. Bio-control and IPM are not right for everyone. As discussed above, the right attitude or mind-set plays a role in successful bio-control and IPM implementation. The differences are quite profound as explained below:

The reactive mind-set, the one with the spray mentality, waits for a pest to make itself known. You see, the reactive mind-set has no scouting program; a customer first screams-bloody-murder about some whiteflies on their selected plants, and only then are the pests discovered. The reactive mind-set attacks the problem in due time with a broad-spectrum, high-potency pesticide; not attacking just the spot infestation, but dousing the entire range — and then some. What's that mix ratio again? 10:1 or is it 20:1, better go with 10:1 just to be safe. The reactive mind-set enjoys little rest, it is often battling those whiteflies, or some other pest.

The preemptive mind-set sees. It knows of the impending whitefly flare-up (it has happened for the past two years). The preemptive mind wants to preserve the sanctity of the chemical-free greenhouse and does so by making small, inexpensive releases of a popular whitefly parasitoid. The preemptive mind acts early. The whiteflies are successfully held at bay for five weeks. [Five weeks during which chemical substances would have normally been sprayed. Perhaps, it saved five spray applications at a cost of $213.00, plus the twelve-and-a-half man-hours which would have been needed to apply the products, and none of which were for scouting. On the other hand, the parasitoids cost the preemptive mind around $285.00 and a total of six man-hours, including the weekly scouting regimen, which should already be an allotment of time considered by our sprayer.] The preemptive mind, with its weekly scouting regimen, uncovers the first tiny flare-up of whiteflies long before a customer could, or would, ever notice. Some parasitism is noted, too. A tiny spot-treatment or the release of a specific predator should do the trick. The cost is still minimal, because the preemptive mind is the preemptive mind and has been able to react before the problem escalated. The whiteflies are managed. The preemptive mind is overjoyed. The preemptive mind rests well at night.

The aforementioned differences aren't normally so blatantly defined, but then again, the general point should be clear enough.

Patience. The fun with chemicals is that they kill so obviously and so quickly (when they work). The results are surefire and undeniable, mostly, anyway. With bio-control though, the results take somewhat longer. Sure, we've heard the odd-but-true tales of swift and sure predatory mealybug-eating beetles going right to work upon release in the interiorscape and having the entire pest population wiped off the face of the earth in only one-and-a-half weeks. (Usually, more time is needed.) However, even at one-and-a-half-weeks, someone's nerves will be shot; the Type-A personality could end up in a padded room in that amount of time.

Patience is a requirement of bio-control and IPM. A necessity. A triumph when successfully dealt with. Without it, bio-control cannot work. Mother Nature needs time to do Her thing.

Attitude. We already know that we must ditch the spray mentality, but there's more to the grower's attitude profile than that. The attitude starts not at the greenhouse or field as you may think, but at your local grocery store with the tables turned — you are now the customer:

Pick up the ripe apple with the small blemish, which is otherwise perfect, and put it in your cart, buy it, then eat it. That's it. You've passed the test. You've allowed Mother Nature into your mind. You've tolerated Her apple's small and inconsequential blemish by putting things into perspective: the blemish was small; it was an otherwise good- and healthy-looking apple. And as far as taste, the apple was by far the pick of the bushel — which wouldn't have been realized if the act of purchasing had been illogically reversed due to a slight, yet natural, imperfection standing in the path of reason.

Now that this is understood, un-turn the tables and began to sell products of your own with that special Mother Nature Touch. Not ill-cared-for plants. Not diseased material. Not malnourished, leggy, yellowed, grossly infested or otherwise poor quality stock; these traits should not be tolerated by anyone. Sell healthy plants. And while you're at it, sell the idea of acceptance. Help your customers understand. There are many things which should be accepted, yet are rejected by the squeamish and the narrow-of-mind. Signs of parasitism and consumption are good things, and should be welcomed as such. If you need help convincing others, no matter how large a customer it is, it never hurts to mention that the plant material in question doesn't have pesticide residue on its leaves, it was grown naturally, it is very healthy. Ask them to just look at the plants and stop scrutinizing them and making judgements which are not necessarily warranted. Here's an exaggerated example of some point-of-sale parlance:

The Buyer: "These plants are covered with aphids."

The Grower: "No, these plants used to be covered with aphids. But those aren't aphids, those are parasitized aphid mummies from which more aphid parasites will hatch."

The Facts: The plants are very healthy. The once substantial aphid population has been dealt a severe blow by the mini-wasps. Many have fled, and 97% of those remaining have been attacked. The damage caused by the aphids will be radically diminished by the time the crop is retail-ready. The buyer of these plants should be happy to see such healthy plants maintained so well with natural enemies. The buyer would do well to simply monitor the active 3% aphid count instead of spraying them. Let the parasitoids keep them in check. Don't poison those valuable wasps for a mere 3% population.

Mummified aphid on bean leaf.

Photo by M. Cherim

The Solution: *Teach your buyer-customer the basics about the predator/prey relationship. Explain what you're doing and how it works. Lend him/her some of your educational resources. Give out The Green Spot's phone number — we'll try our best to help.*

Moneythink. *Now here's a subject of great obscurity. Determining before-hand what your first-year bio-control and IPM costs are going to be is like knowing what your first-year chemical costs are going to be. You can figure, guess and estimate (based on the application rates and pricing information shown in this manual) what you can expect to be doling out if you know what kind of pest pressure to expect.*

Without that information, though, expect the following: Initially bio-control and IPM practices may seem, and in many cases might actually be (5-45%), more expensive than conventional means. Initially. Rumor has it, circulating among The Green Spot's contacts, that long-term costs are expected to be in-the-ball-park with conventional methods, or less. (A lot may depend on how well your customers are taught.) To help clear the confusion, below are listed a few industry examples with a brief summary added for each.

Bedding plant growers: *Expect to pay more for bio-controls unless your clientele is really easygoing and understanding. Spring will be your greatest season of expenditures on high volume and frequent releases. If you factor in short-term labor savings, restricted entry intervals or REIs, plant quality and losses, positive community press and fewer legislative headaches, you may consider yourself in an enviable position.*

Ornamental, pot plant and herb growers: *Ditto. What holds true for bedding plant growers holds true for this group, too. One exception is the time of year: the greatest season of expenditures and the frequency of releases will be less defined, and the releases should be of lower volume over a longer period, unless the crop in question is a seasonal one.*

Vegetable growers: *Vegetable crops can usually tolerate more pest pressure than other crops. Not to say it is easy or that pests can have the run of the place, but they may be allowed a greater foothold. Additionally, disease management must be considered, as does crop health and vigor, and the quality and quantity of the yield. The crop-cycle will determine the greatest expenditure period with its frequent, but small, releases. If you factor in short-term labor savings, REIs, crop and yield quality, fewer legislative headaches, you could be well ahead of the crowd. Since vegetable growers typically have a near-monoculture, scouting, record-keeping and IPM can become very organized. Most vegetable growers can readily save money over conventional techniques if the methods are appropriately applied. It may be a good idea to check out the organic market. Perhaps, it could prove to be a viable revenue stream, if your product qualifies.*

Field/row farmers, nurserymen and gardeners: *On the mega-scale, bio-control will *usually be more expensive than conventional methods. The economies-of-scale only go so far in the bug-biz. Fortunately, on any outdoor scale, natural, indigenous predators and parasitoids, will normally be present in a no-spray crop. (This is also true of many open-system interior and greenhouse sites.) The releases you'll usually partake in will be used to supplement the existing population of natural enemies. It'll tip the balance temporarily in your favor —Mother Nature will typically straighten things out in due time, though. The period of the greatest expenditures will be during the beginning and end of the crop cycle. (*A couple of expense exceptions: Our largest field farmer has 12,000+/- acres of mint. He is successfully controlling spider mites and saving money. Another customer, a six acre cabbage grower, actually saved over $1,000.00 controlling aphids and diamondback moth larvae biologically. Moreover, his results were absolutely astonishing.)*

Interiorscapers: *The trend seems to be repeated, low-volume inoculations of various predators and parasitoids. This practice is bringing about better results than can be achieved with sprays, but at a price that is slightly to moderately more. Releases need to be made monthly, or in that neighborhood, so figure on adding $X to your next bid. $X should be based on the size of the account. Interiorscapers are sharing stories of triumph. The money being spent seems to be acceptable because the results are there and the interiorscape contractor is already so limited in what's available to him/her to use. Considering this, the Green Methods will cost more in some interiors: $X on bio-controls is greater than zero spent on chemicals which don't yet exist; in return, however, you get solid pest control.*

Landscapers: *It's hard to say where this group fits in, financially speaking. If the typical project entails the eradication of lawn grubs, the outlay of funds won't be too great. Same goes for the typical aphid or spider mite intervention, say, on roses. The season of the greatest expenditures will probably be the summer months, with emphasis in the early summer [late spring] and late summer. If it is grub control, or aphid or spider mite suppression, the costs will probably be minimal and, more than likely, the results will be enviable.*

But don't get too chipper, yet. It may sound good, but things can get rough [with bio-control, IPM or chemicals] if you happen to be dealing with some scale insects in a tree, or something with similar potential. It can be done, but the costs can get high.

Everyone: *Many variables will dictate how much money or what kind of budget will be needed. Your geographic location and its proximity to your preferred bio-control/IPM supplier and the associated freight costs will play a partial role. For example, New England customers can save a lot on freight when making their purchases from The Green Spot because the items can be shipped via ground methods. Volume buying can help reduce unit costs; the economies-of-scale do play* a small part. The Green Spot offers some ways to save. Give us a call or see the applicable section(s) of this manual.

Other variables will also play a part. They are actually too numerous to mention them all. Some, however, are: pest pressure, scouting effectiveness, beneficial organism effectiveness, implementation of IPM resources, self education, consumer education (getting better by the day), timing, temperature, light-levels, photoperiod, humidity, host material, environmental setting, luck, weather, crop type, crop health, etc. You get the idea, there are many.

All in all, Bio-control and IPM seem to work at a level which is acceptable to most keepers-of-the-green. The general consensus is that it costs more, but the few extra dollars are worth it because, as it seems, no one, not growers, not farmers, not interiorscapers, likes to spray and would rather have a fairly economical means which works comparatively well instead: hence, the Green Methods.

We stay in touch with our contacts to see what's working and what's not. We appreciate the feedback, both good and not-so-good. And from the gist of what we're hearing, bio-control and IPM are pretty good things to hang in your pest control gun-belt. ▩

Our great grandparents used bio-controls...

They just didn't know it.

time to proceed

This subsection will hopefully prepare you for using the next main section which details the individual critters. Here you can understand what is expected of the grower who initiates a bio-control/IPM program. While some folks who control pests can do it like the reactionary spray man, the IPM set has to do things differently. IPMers must have a lot of important information readily available so as to not only know when to act, but how to act, as well. (This is described in greater detail earlier in the Green Methods 101 section.)

Starting out. One way to get help, if you don't want to go it entirely on your own, is to ask your bio-control and/or IPM supplier for their assistance — they should offer verbal consultation over the telephone at no charge (though asking for detailed, written guidance via e-mail or fax will probably require significantly more time and may be considered a billable item; so have a pencil and paper ready when you call.)

Like The Green Spot, most suppliers have a vast network of users from which they constantly draw information, knowledge and experience, and will try their level best to extrapolate important details from your available information and make a firm, honest program recommendation geared towards success. (Though you would probably like to have the supplier gear the program solely towards saving money, an honest bio-control/IPM program should not be structured that way as it will, perhaps, be a detriment to your program's efficacy.)

The following is a list of facts and specs your bio-control and/or IPM supplier will (or should) ask before being able to make an accurate recommendation:

Typical questions

1) What crop or crops are being grown?

2) How is it being grown and where?

3) What is the size of the growing area or structure?

4) What pest(s) are you dealing with?

Hypothetical answers

1) Pothos.

2) In a soilless media filled planter bed in a interior bank lobby.

3) 14,000 square feet.

4) Citrus mealybugs.

Typical questions cont...	*Hypothetical answers cont...*

5) Are these pests in an area greater than or less than 70% of the growing area or structure described in answer #3?

5) Less than.

6) Because the answer to #5 was less than, what percentage of the structure is affected? (If greater than, the entire structure's size is used.)

6) About 5% (700 square feet).

7) If the foliage portion of the plants exceeds 3 feet in height, how tall is it? (If over 3 feet tall, every three foot or less section of vertical height would be counted as a "level." Two levels would double the area's actual size.

7) They're under 3 feet tall (1 level).

8) To what degree are the pests present?

8) The most recent count was approximately 15 mealybugs per square foot... a high infestation by our standards.

After compiling this information, you or your consultant can do some simple math to determine what's needed. Keep reading to learn how The Green Spot's application formulas work. Then try to come up with a recommendation for the situation described above. Check your work against the solution we've come up with at the end of this section to see if you're ready to start.

Determining usage rates. In 1997 we refined and tested all of our application rates in order to better facilitate real-world scenarios. Our objective was to make release rates even more exact and methodical so that you could save money without sacrificing what we felt should be called a "safe-rate threshold" (the least number of organisms needed to perform the duties intended). Based on the results we gathered, the feedback we received, and the ease of use we experienced while helping others, we feel we hit our target right on the mark.

One difficulty we ran into in prior to 1997 was that in order to be more accurate, we needed be more incremental in our prescriptions for use. We solved that problem by developing a new "Rate Chart." The prototype we designed [for Cryptolaemus montrouzieri (a mealybug predator) as an example], is shown on the next page, fig. 2.

This release rate chart has many levels of application: **preventive releases** *(made before the first applicable pest is seen);* **low infestation releases** *(made at the first sign of the applicable pest);* **medium infestation releases** *(made when higher numbers are present);* **high infestation releases** *(made when the infestation is full blown);* **maintenance releases** *(made when things are seemingly under full control and all that is required is to keep things clean).*

The second difficulty was to determine, by pest, how many individuals in a given area would be needed to consider the infestation low, medium or high. We decided, with few exceptions, that it was basically impossible to make the desired determination since information of that variety is normally fueled by personal opinion and the location in question.

RATE CHART EXAMPLE   Fig. 2
For legend to abbreviations, see below...

> PRVNT: N/A
> CTRL-L: 2-4 / *YD. - TRI-WKLY - 2-3X
> CTRL-M: 4-6 / *YD. - TRI-WKLY - 2-3X
> CTRL-H: 6-8 / *YD. - BI-WKLY - 2-4X
> MAINT: 2-3 / *YD. - MTHLY - INDEF
> GRDN: 20-60% of rates above.
> ACRE+: 3-20% of rates above.
> COMMENTS: Large scale use is normally in
> the southern states. Usually in Citrus.

LEGEND:

PRVNT = Preventive releases.
CTRL-L = Low infestation control.
CTRL-M = Medium infestation control.
CTRL-H = High infestation control.
MAINT = Maintenance releases.
GRDN = Garden. Small scale outdoor use.
ACRE+ = Large scale outdoor use.
***YD.** = Yard (either cubic or square).
WKLY = Every week.
BI-WKLY = Every other week.
TRI-WKLY = Every third week.
MTHLY = Monthly.
QTRLY = Quarterly.
NEED = As needed or until change.
INDEF = Indefinitely or until change.
?X = Number of times.

MEASURING INFORMATION:

<70% planted/infested = If the area or structure is under 70% planted/infested, then use the square or cubic yardage of only the actual area in question to calculate how many organisms to use. This rule applies to all indoor and outdoor plantings.

>70% planted/infested = If the area or structure is at least 70% planted/infested, then use the square or cubic yardage of the entire area or structure to calculate how many organisms to use. This rule applies to all indoor and outdoor plantings.

<3' tall plantings = For plantings which are primarily under 3 feet in height, determine the square yardage of the structure or area, or planted/infested area (refer to the rules above), to calculate the appropriate application rate. This rule applies to indoor and outdoor plantings.

>3' tall plantings = For plantings which are primarily over 3 feet in height, determine the cubic yardage of the structure or area, or planted/infested area (refer to the rules above), to calculate the appropriate application rate. If it is easier, try doing the regular square yardage calculations, as described in the paragraph above, then add 100% of the determined rate for every three feet of plant height. A peculiarity of tall plantings, though, is that many are mostly trunk, i.e., many palm tree species. In cases like that, calculate the foliage area alone (unless the trunk is affected). This rule applies to indoor and outdoor plantings.

Mixed height plantings = For plantings of mixed heights, do the best you can, applying all of the applicable rules above. If it helps, sketch a scaled 3-D elevation detail of the area and "cube" the applicable areas, or fax us your sketch, then call us.

Outdoor only rates = First, due to the presence of indigenous natural enemies, it is usually necessary to only use a percentage of the recommended rates. What percentage, however, will have to be determined by looking at your scouting reports. This is also true of small plot and garden spaces. Second, for the sake of mathematical ease, we will say that an acre is 4500 sq. yds. instead of 4840 sq. yds.

Help! = We do these calculations all the time. Measure your area, than give us a call. We can easily determine what you'll need to use.

Other considerations = Every control rate has a range of numbers. This range is useful in being more precise, i.e, low-end of high infestation versus high-end of a high infestation. Extreme plant [foliage] density may also be a reason to use the upper end of these ranges. Experiment and determine what exactly is best for your situation.

Metric applications = Add approximately 10% to the square yardage rates for square/cubic meter rates. ▨

The solution. Are you ready to rate your own? Let's find out. Read on and see if your answer to the hypothetical situation matches ours. Doing some simple math, utilizing the current, predetermined formulas (shown on previous page), certain figures can be obtained. Let's review the facts and look at this situation one piece at a time:

1) 5% or 700 sq. ft. of a 14,000 sq. ft. bank lobby has pothos groundcover which is infested with citrus mealybugs.

2) To determine the yardage, divide 700 sq. ft. by 9 (there are 9 sq. ft. in a sq. yd.). The answer is 77.78 yards. (If other than low-growing pothos, the foliage height of the plants might have to be considered. If this were the situation, one would multiply the 700 sq. ft. area by the number of 3 foot levels, before dividing by 9; this will give the cubic yardage.)

3) Crypts, which are the best choice of predator for citrus mealybugs, would have to applied at the rate of 6-8 beetles per yard to counter a high infestation. Since conditions are pretty straightforward, we'll start at 7 beetles per yard. Multiply 7 by 77.78 yards and you'll come with 544.46 beetles. Since the closest standard product size is 500 beetles (this can be a drawback), that's what we would recommend. Moreover, based upon the rate chart, repeated releases are prescribed. In this case, make releases every other week, 2 to 4 times. We would prescribe a two-shot standing order of 500 beetles, with instruction to scout the infestation to see if additional applications are needed at orders' end.

If you came up with roughly the same answer, you're probably ready to proceed on your own. Check your work with your supplier, though. Changes in formula, special considerations (using two or more organisms for the same purpose, etc.), may apply. Additionally, since The Green Spot's formulas are not yet standardized throughout the industry, your favorite supplier may totally disagree with our formulas and want to go about it completely on their own. Realistically, though, their recommendation should be pretty darn close to ours. If the supplier fails to suggest multiple releases, or denies their necessity, get a new supplier.

The transition. *Okay, now you're ready in every conceivable way: you've got your head screwed on right, you're scouting like a seasoned pro, you've studied, etc... except one, you've been spraying up until now, and according to the biorational substances chart (fig. 1) shown in this book on page 12, the stuff you've been spraying (since it's not listed) might be harmful to the bio-control agents you want to employ. Now what?*

Call your supplier, having ready the chemical name of the product(s) you've been applying, including fungicides, plant growth regulators, etc., if we've just described your situation. Many of the bio-control suppliers around the world will be able to help you in determining how much time you should wait before making your first bug release. Since a considerable amount of time may have to pass, depending upon what you've been using, it is advisable to ask these question up to 90 days prior to the day you wish to make your initial release.

Your favorite bio-control supplier will gladly help you, not only with information on residual periods for chemicals, but they'll give you menu selections of what methods and/or products to use in the interim — as the need to control pests probably won't go away solely because of your good intentions.

The transitions from chemicals to the Green Methods can be a slow and arduous process. But it doesn't have to be if you use only half of what's available to you. Typically, based on actual accounts with our clients, the first year will bring a mix of wins and losses — mostly wins — and you'll probably spend a little more money than you hoped. In year two, things will begin to come into focus; you'll still get assistance from your supplier, but you probably won't need as much. Year three is a common fledgling year: growers go it on their own but still call when trouble arises. By the fourth year, from what we've noticed, our clients will fax their standing order for most of the year. And during the year we'll speak only one to two times.

It is doable, just be prepared to think and to use what's available to you. Use your imagination; sometimes the simplest things will prove to be the most useful. And whatever you do, don't quit because of an early setback or two. Setbacks are a normal part of the learning process. As a reminder, our clients have an average 85% success rate when they follow the rules.

Are chemicals better? Probably not. We hear all kinds of grievances about chemicals. More often than not, though, the chemicals don't work because they are not applied correctly. Ditto for the bugs and IPM tools. Follow the manufacturers' instructions on whatever products you use. That ensures the safest and best results in nearly all cases.

Call The Green Spot, Ltd. at 603/942-8925 for assistance if you want. ✖

Gallery

of

Bugs

*a close-up look at
the good, the bad
& the ugly*

photo, opposite, by D. Simser

aphids

Order: *Hemiptera* Sub Order: *Homoptera* Superfamily: *Aphidoidea*

Aphids are probably the most common pest we hear about, overall. Besides being called aphids, these sucking plant pests sometimes called plant lice; more often than not, though, growers tell us about the %*#&$!! infesting their crops.

Aphids tend to be very communal (as shown in the photo, inset), often found in large congregations attacking the tender parts of plants, particularly the new growth shoots and tips. Equipped with a sharp,

Aphid banded together, killing a leaf.

straw-like proboscis, aphids penetrate the plants' cells, removing the sap. The sap is converted to energy absorbed by the pests, then expelled as a sugary, sticky fecal substance aptly called honeydew. Watch out for this stuff, ants like it (see photo, below) and will protect aphids — as well as other pests — for it, plus it promotes the growth of black sooty mold (an unsightly mold which grows on the agar-like honeydew).

The honeydew-eating ant: bio-control's worst foe.

Both photos by M. Cherim

Usually the first sign of aphids will be winged adults (see photo, facing page, top) in the spring. This is because the aphids have come from another area or host. [Another time winged adults may be seen is when they become overcrowded; some aphids will be develop wings to accommodate

*flying off to expand their range or to seek new host material for overwintering.]
Some aphids utilize different host plants — typically laying eggs on woody plants
— in the winter months, when sexual reproduction takes place. In the summer
months, their reproduction is asexual, giving birth to live, complete aphids, ready to
feed. Their population growth during the spring and summer can be deemed
exponential. They can fully develop in mere days if conditions are right.*

*In a greenhouse, however, asexually reproducing aphids can be present year
'round. No wonder they strike fear in the hearts of greenhouse or structure growers
more than any one else.*

The winged aphid. Where trouble begins.

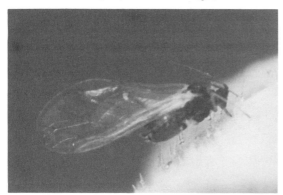

*Finding aphids early,
or acting before they
even show up, is, by
far, the safest way to
deal with these
notorious pests. Of
course, proper
scouting is the key to
spotting and manag-
ing an early flare-up.
The early pre-
introduction of aphid
parasites is also highly*
recommended.

*The scout should always look for aphids on the plants' most tender new growth.
Other signs to be on the lookout for are: white skins shed by the developing aphids
as they molt through their juvenile instars or stages, which will be present directly
below the aphid mass and; ants which might be seen marching up and down the*

*plants' stems as they
transport their newly
harvested honeydew
booty back to their
mound or nest.*

*For more detailed
information and
recommendations
concerning aphids and
ways to control them,
you are invited to call
The Green Spot, Ltd.
at 603/942-8925.* ✄

The progeny. How trouble continues.

Photo, top, by P. langlois, courtesy of Applied Bio-nomics, Ltd.
Photo, bottom, by M. Herbut, courtesy of Applied Bio-nomics, Ltd.

aphid controls

The Aphid Parasitoids

Aphidius spp.
colemani and matricariae
(ah-FID-ee-us kole-eh-MAH-nee, may-tree-KARRY-ay)

Description— These 2-3 mm. mini-wasps are best used for preventing the establishment of **more than 40 species of aphids**. They can also tackle minor to medium infestations. And, if established, they can adequately protect a crop throughout the season, with some possible exceptions in late summer (see Drawbacks).

A. colemani, which are shipped as ready-to-hatch mummies, seem to be the product of choice when **melon or cotton aphids (*Aphis gossypii*)** are present.

A. matricariae, which are shipped as pre-hatched, pre-fed adults, are the product of choice when aphids other than *A. gossypii* are present. *Some* popular hosts of this mini-wasp include: the **bean aphid (*Aphis fabae*)**, (see photo, facing page; the **potato aphid (*Macrosiphum euphorbia*)**; the **glasshouse potato aphid (*Aulacorthum solani*)**; the **pea aphid (*Acyrthosiphon pisum*)**; and the **green peach aphid (*Myzus persicae*)**, probably the most common aphid.

Aphidius sp. with ovipositor at work (through the forelegs).

Photo by M. Badgley, courtesy of Buena Biosystems, Inc.

Life-style— The *Aphidius* spp., as parasitoids, work by laying eggs in the abdomens of aphids (see photo, left). And they can lay 200-300 eggs! The wasps' larvae which hatch from the eggs, slowly weaken and kill the aphids from within (*endo*parasitism). The aphids then turn into "mummies" (see photo, page 13) as the wasps pupate.

The life-span of these parasitoids is roughly 2 weeks in their immature stages, then 2 weeks as adults. The conditions for optimum performance will be between 64-75°F with a relative humidity of around 80%. But these are optimum conditions, and not necessarily a prerequisite of successful implementation. Please note, however, cooler temperatures will hamper reproduction and development a certain degree.

Bean aphids attacking nasturtium.

Benefits— These wasps will work in fairly cool areas with low light levels and short photoperiods. Moreover, they are really easy to scout (see Scouting).

Aphidius spp. wasps are superb preventive agents, thus offering growers a potential money-saving tool. Additionally, they can establish themselves in nearly any region of the country; they overwinter in the toughest climes. Once established, growers might be able to reduce the size of releases made due to the presence of on-site wasps: another money-saver.

A. matricariae, because they are shipped as adults, offer the benefits of a guaranteed sex ratio (95% fem.), fast oviposition (or egg-laying, we've seen it happen in as little as 10 mins.) and guaranteed freedom from hyperparasites (see Drawbacks).

Drawbacks— One thing to consider when using the *Aphidius* spp. is the potential of hyperparasites. These even smaller wasps parasitize the *Aphidius* spp. while in the aphid. So instead of another *Aphidius* spp. hatching out, the hyperparasite will emerge in its place. This can be very disruptive to your bio-control program.

For this reason, ordering adult parasitoids when possible is recommended, especially in late summer when the *A. colemani* might be contaminated. Please note, however, it has been suggested that too much emphasis might be placed on the potential of getting hyperparasites. Due to modern production processes, getting hyperparasites with your *A. colemani* order is extremely rare. Once they're on site, though, anything can happen

Nevertheless, your *A. matricariae* can develop this problem once released at your site, too, especially in the late summer (see Scouting).

In addition to the rare problem of *A. colemani* being shipped with hyperpara-

 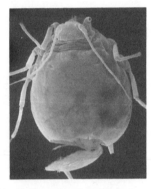

sites, because they are mummies, their sex ratio and hatch rate cannot be accurately predetermined, plus they must feed and mate prior to oviposition. The latter may cause a lag in the performance schedule.

The exit hole of the Aphidius spp. (left) is smooth and clean; the exit hole of the hyperparasite (right) is jagged and uneven, and the flap remains.

Scouting may be somewhat hampered because the use of yellow sticky traps as a monitoring tool may be hampered (see Advisories).

Scouting— To determine if your mummies are hatching out more *Aphidius* spp. wasps or if hyperparasites are emerging, take a close look at the exit hole. The emergence of the *Aphidius* spp. wasp produces a clean, round hole without jagged edges. And often the flap or lid of removed material is absent (see the comparison photos which follow below).

Scouts have the obvious mummies and exit holes to look for, but with these agents, there will probably be some visible and nearly instant reduction in the pest count. When *Aphidius* spp. are released, aphids often send a scent signal of alarm. We've actually seen them falling from a plant trying to flee.

An additional sign of parasitism — early parasitism — and *Aphidius* spp. activity in your crop is the tiny dark-orange to reddish/brown oviposition [sting] mark which may be present on the back-end to top of the aphids' abdomens.

Just prior to the wasp pupating, its host aphid will turn a greyish color. This, however, depends upon the host species.

Both S.E.M. photos, this page, by N. Cherim
Photo, facing page, by M. Cherim

Advisories— Yellow sticky traps should be removed prior to releasing these mini-wasps. To monitor for thrips, use blue traps. If yellow traps must be used for whiteflies, etc., hang them for only two days per week.

Ants, if present, should be controlled. They will defend aphids from predators and parasites to protect their honeydew food (the excrement of aphids, yuk!). Use barrier, exclusion products or boric acid products to control the ants.

If your planting doesn't have any ants, check to be sure that the honeydew isn't too heavy. This may

A. matricariae unit of 500 adults.

prove to be a hindrance to the parasitoids' performance; they may spend too much time cleaning themselves.

Usages— Greenhouses, fields, interiorscapes, orchards and gardens. We've seen the successful preventive and curative implementation of these species in just about every conceivable situation.

Rates—

```
PRVNT: 1-2 / YD. - BI- WKLY - NEED
CTRL-L: 2-4 / YD. - WKLY - 2-3X
CTRL-M: 4-8 / YD. - WKLY - 2-4X
CTRL-H: N/A
MAINT: 2-3 / YD. - MTHLY - INDEF
GRDN: 40-80% of rates above.
ACRE+: 10-30% of rates above.
COMMENTS: For A. colemani, ADD 20% to
the rates above or make 1 additional release.
```

Pricing information— The *Aphidius* spp. covered in this section can be obtained from The Green Spot by calling 603/942-8925. Detailed release instructions are provided with every order. Here is the current, industry-average pricing for 1998...

Item no. CAC5C ... 500 A. colemani mummies ... 1 unit = $26.32

Item no. CAC5C-6 ... 500 A. colemani mummies ... 6-11 units = $24.87 ea.

Item no. CAC5C-12 ... 500 A. colemani mummies ...12+ units = $22.38 ea.

Item no. CAM5C ... 500 A. matricariae adults ... 1 unit = $29.58

Item no. CAM5C-6 ... 500 A. matricariae adults ... 6-11 units = $26.92 ea.

Item no. CAM5C-12 ... 500 A. matricariae adults ...12+ units = $24.76 ea.

The Aphid Midge

Aphidoletes aphidimyza
(ah-FID-o-lee-teez ah-FID-ih-my-zuh)

Description— The predacious larvae of these delicate, 3 mm. mosquito-like midges (see photo, facing page) are best used for controlling substantial populations of more than **60 species of aphids, including those mentioned for *A. colemani* and *matricariae*.** If established, they can adequately protect a crop throughout the season.

A. *aphidimyza* adults are very nomadic — travelling to where the action is — decimating populations of even some obscure aphid species.

The midges are shipped as

The tub that A. aphidimyza is shipped in makes an ideal hatching unit. The one above contains 1000 puparia.

puparia (the last immature stage) in a vermiculite medium. The tubs they are shipped in serve well as hatching tubs (see photo, above).

Life-style— The *A. aphidimyza* female adults, being nomadic, as mentioned above, actively seek out colonizations of aphids, like that shown to the lower left. Mated females, when they find these groupings, will lay eggs amongst the aphids — up to 250 of them. The eggs hatch into orange larvae which grow to 3 mm. (See photo, inset)

These larvae kill

The A. aphidimyza larva shown is attacking a winged aphid adult.

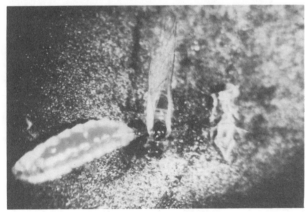

Photo, top, by M. Cherim
Photos, bottom and facing page, by P. langlois, courtesy of Applied Bio-nomics, Ltd.

aphids by injecting a toxin through their legs. They then eat the aphids or move on. Either way, the aphids die. Doing this, *A. aphidimyza* can destroy up to 50 aphids per day!

The life-span of these predators is roughly 4 weeks in their immature stages, then less than 2 weeks as adults. The conditions for optimum performance will be between 64-77°F with a relative humidity of

This delicate, mosquito-like midge is tough as nails to over sixty different species of aphids.

around 70%. But these are *optimum* conditions, and not necessarily a prerequisite of successful implementation. Please note, however, cooler temperatures will hamper reproduction and development a certain degree.

Benefits— If released into high infestations, *A. aphidimyza* larvae tend to kill much more than they'll eat. This makes them capable of handling heavy pest pressure.

These midges are extremely versatile; they are successfully used in many environments, including street trees, orchards, interiors, etc. They're able to search in very high places. One of our customers successfully used them in 60 foot pecan trees in south-central Texas.

Drawbacks— The availability of *A. aphidimyza* is excellent, and it over-winters well all over the country, but it undergoes diapause (a quiescent state, hibernation) when temperatures dip below 40°F for extended periods or the photoperiod is less than 12 hours (D).

Another drawback with this predator is the difficulty in finding it (see Scouting).

Additional consideration should be given if pesticides are part of your IPM program; these midges are very sensitive (see Fig. 1 on page 3).

Scouting— These predators are hard to scout. The best thing to look for is adult presence. They may be seen during overcast weather or in the evening flying. Sometimes you can glimpse their mating dance, in which great numbers will be hovering in a tight cloud of bodies. This cloud will appear to be suspended in midair.

The larvae may be seen at times. Since they are orange, they do contrast quite well with most foliage.

Another sign of their activity is the high numbers of decimated aphids which can be left in their wake, plus a reduction of live aphids.

Advisories— To counteract the natural urge for these predators to undergo diapause you must 1) keep the temperature above forty and, 2) provide supplemental lighting during the appropriate time of year. A 60 watt bulb for every 60 foot radius will do the trick. Bear in mind, however, most organisms, regardless of nature, will normally slow down a degree in the winter months.

Ants, if present, should be controlled. They will defend aphids from predators and parasites to protect their honeydew food (as it is with the *Aphidius* spp.). (See photo, left.) And this is *very* critical with *A. aphidimyza*. But at least heavy honeydew doesn't bother these predators as much as it does with others. In fact the adult midges will sometimes feed on the 'dew themselves.

Ant harvesting honeydew from aphids.

A. aphidimyza's larvae drop to the ground or medium to pupate. And because of this your site must be adequate to provide for that activity: It must have a friable medium, or something that they can *enter*, sod is okay. Moreover, it is advisable to not use these predators on lettuce as the larvae might possibly drop between the leaves when they pupate. Hydroponic systems might also be inappropriate for the same reason.

Additionally, it is probably a good idea to halt the use of *parasitic nematodes — they will attack some of the pupae. If you are going to use *A. aphidimyza*, first determine that your site can foster their reproductive habits. (*Try using *Hypoaspis miles* for fungus gnats instead (see thrips controls.)

Usages— Greenhouses, street trees, fields, interiorscapes, orchards and

Photo by M. Cherim

gardens. We've seen the successful implementation of this species in just about every conceivable situation.

Rates—

PRVNT: 1-3 / YD. - MTHLY - NEED
CTRL-L: 2-5 / YD. - BI-WKLY - 2-3X
CTRL-M: 4-7 / YD. - WKLY - 2-4X
CTRL-H: 6-9 / YD. - WKLY - 3-5X
MAINT: 2-3 / YD. - MTHLY - INDEF
GRDN: 50-80% of rates above.
ACRE+: 20-60% of rates above.
COMMENTS: Preventive releases should be made in sites open to the outside.

Pricing information—
The *A. aphidimyza* midge covered in this section can be obtained from The Green Spot by calling 603/942-8925. Detailed release instructions are provided with every order. Here is the current, industry-average pricing for 1998...

Item no. CAA250 ... 250 A. aphidimyza pupae ... 1 unit = $12.00

Item no. CAA250-6 ... 250 A. aphidimyza pupae ... 6 + units = $10.20 ea.

Item no. CAA1M ... 1000 A. aphidimyza pupae ... 1 unit = $23.10

Item no. CAA1M-6 ... 1000 A. aphidimyza pupae ... 6-11 units = $21.94 ea.

Item no. CAA1M-12 ... 1000 A. aphidimyza pupae ... 12+ units = $20.84 ea.

A Green Lacewing Mix

Chrysoperla spp.
carnea, comanche & rufilabris
(kry-SOPE-er-lah KAR-nee-ah, KOH-man-chay, ru-fil-LAY-briss)

Description— These nocturnal predators come in three major forms: eggs, larvae and adults. The eggs are useful when you're in no great hurry to get rid of the pests. The larvae are useful for the quick cleanup. And the adults, being nomadic like *A. aphidimyza*, are useful in tree applications.

The larvae are the only predatory form of this insect. And what a form they are; very opportunistic. Like the two preceding bio-control agents, *Chrysoperla* spp. **can tackle a great number of aphid species**. Moreover these predators may eat outside of their aphid-preference diet to enjoy **other soft-bodied pests: scale insect immatures, including long-tailed and other mealybug species; whiteflies and others, especially certain insect eggs.**

The eggs are shipped loose in an inert medium of rice-hulls. The rice-hulls are a distribution carrier to facilitate the proper placement of the eggs.

The larvae are very cannibalistic and must be separated in transit. This is accomplished by means of a frame or hexcell unit (see photo, inset). This unit is comprised of little compartments which can be opened a-row-at-a-time for predator release.

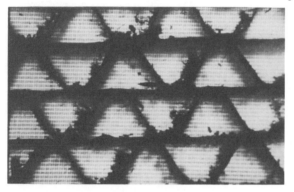

The adults come in a tube screened at both ends. Often-times they are already laying eggs inside the tube.

Holding the hexcell unit up to the light is a good way to check on the viability of the larvae.

Life-style— The 2 cm. *Chrysoperla* spp. female adults, being nomadic, as mentioned above, actively seek out colonizations of aphids. Mated females, when they find these groupings, will lay their 1 mm. light green eggs which are perched atop 1 cm. long filaments, amongst the aphids — up to 200 of them. The eggs hatch into tan-colored, alligator-like larvae (see photo, facing page), which grow to 8 mm., and are extremely voracious feeders which will go right to work on the aphids — and each other. They can consume 100 aphids or more!

The life-span of these predators is roughly 30 days in their immature stages, then less than 2 months as adults (see photo, below). The predacious larval stage lasts roughly 15-20 days. The conditions for optimum performance will be between 67-89°F with a relative humidity of 30% or greater. But these are *optimum* conditions, and not necessarily a prerequisite of successful implementation. Please note, however, cooler

The lacewing adult's large eyes give it excellent dusk/dawn vision — times when it is most active.

Photo, top, by D. Simser
Photos, bottom and facing page, by M. Badgley, courtesy of Buena Biosystems, Inc.

temperatures will hamper reproduction and development a certain degree.

Benefits— Cost. The eggs are a fairly economical method of application. However, with the eggs, you sacrifice some effectiveness and speed. For those traits, the larvae, or "aphid lions" as they're sometimes called, are the way to go.

The larvae are one of the fastest predators we have, from release to first meal, anyway. Moreover, because of their opportunistic nature, they are useful for many pests in addition to aphids. For reliability, though, use them for aphids and, perhaps, scale insect species.

A voracious lacewing larva attacking an aphid.

The adults are effective when treating orchards and such. The adults are nomadic as explained previously.

Drawbacks— Well, for one thing, the larvae, especially the larger ones, can deliver a painful little bite (to people and each other). This is not to scare you. It's insignificant compared the benefits. And never have we had negative client feedback regarding this slight drawback. Nor have we heard negative comments regarding the use of this predator in interiorscapes. Except one...

The adults, nocturnal as they are, come out at night. Once, in a hospital cafeteria, the night-shift did complain to one of our interiorscaper-contractor-customers of multiple adult sightings. The problem was short-lived, though.

Another drawback is that they're difficult to scout (see next).

Scouting— The scout is going to be hard-pressed to find actual larvae or adults by day; they are usually well hidden. If the scout wants to locate larvae and adults, he or she should plan on doing so either in the evening or on an overcast day.

Clean new growth is one sign effective as a scouting aid. Another is to look for the eggs. They are usually on the top surfaces of the leaves. Decimated or sucked-dry-looking aphids are another scouting sign.

Advisories— Ants, if present, should be controlled. They will eat lacewing eggs and defend aphids from predators to protect their honeydew food. The ants actually sort of "herd" the aphids as they tend to their needs (see photo, page 32). Use barrier products or boric acid products to control the ants.

Pollen, nectar and even honeydew will help sustain the adults. They are not predacious but do need food. A product such as Biodiet™ (a honeydew substitute discussed in this manual) may prove useful for green lacewing adults. It is upon this type of formula that lacewing adults are commercially reared.

Usages— The eggs and larvae are useful in greenhouses, fields, interior-scapes (though not where the public will be after hours), orchards and gardens. We've seen the successful implementation of these species in just about every conceivable situation. The adults should be used only in row crops, trees, orchards and, possibly, tall interior plantings. The adults will lay eggs next to aphid colonies as discussed previously. Therefore, consider using another bio-control agent for tree pests other than aphids.

Rates—

```
PRVNT: 1-3 / YD. - MTHLY - NEED
CTRL-L: 2-5 / YD. - BI-WKLY - 2-3X
CTRL-M: 4-8 / YD. - BI-WKLY - 2-4X
CTRL-H: 7-12 / YD. - BI-WKLY - 3-5X
MAINT: 1-2 / YD. - TRI-WKLY - INDEF
GRDN: 60-90% of rates above.
ACRE+: 20-50% of rates above.
COMMENTS: Rates shown above for larvae.
For eggs multiply rate times 5, adults divide by 4.
```

Pricing information—
The *Chrysoperla* spp. covered in this section can be obtained from The Green Spot by calling 603/942-8925. Detailed release instructions are provided with every order. Here is the current, industry-average pricing for 1998...

Item no. CCSE5M ... 5000 Chrysoperla spp. eggs ... 1 unit = $17.44

Item no. CCSE5M-8 ... 5000 Chrysoperla spp. eggs ... 8+ units = $15.69 ea.

Item no. CCSE20M ... 20,000 Chrysoperla spp. eggs ... 1 unit = $62.98

Item no. CCSE20M-4 ... 20,000 Chrysoperla spp. eggs ... 4+ units = $56.64 ea.

Item no. CCSL5C ... 500 Chrysoperla spp. larvae ... 1 unit = $25.20

Item no. CCSL5C-6 ... 500 Chrysoperla spp. larvae ... 6+ units = $23.18 ea.

Item no. CCSA5C ... 500 Chrysoperla spp. adults ... 1 unit = $118.30

Item no. CCSA5C-4 ... 500 Chrysoperla spp. adults ... 4+ units = $106.47 ea.

The Mirid Plant Bug

Deraeocoris brevis
(dare-ay-OCK-orr-uss brev-iss)

Description— These vicious predators, named after their family name *Miridae*, were classified in our 1995/96 publications as thrips predators. This is still true, but secondarily so. Experience speaks loudly and, in the case of *D. brevis*, it clearly tells us this "general" predator **prefers aphids and whiteflies to thrips.** And based on one occurrence in 1997, they *really* like aphids: they provided an over-the-weekend cleanup of a *major* aphid infestation.

In addition to the pests mentioned above, these predators like **small larvae (caterpillars, etc.), an array of insect eggs, tarnished plant bugs** (with which recent work has been promising), **and a host of other organisms such as pear psylla** (on which the original culture was feeding when it was first collected).

D. brevis acts as a "general" predator. This is common among "true bugs." True bugs tend to be thorough at whatever they do. In the case of these predators, they can thoroughly clean house.

D. brevis are shipped as adults with some nymphs present. Both the adults and nymphs behave as predators.

Life-style— The large 5 mm. adult female bugs (see photo, inset) lay eggs in plant tissue — up to 200 of them. The nymphs which hatch may taste the plant, but will cause no noticeable damage. By the time the nymphs reach their second instar (growth stage), they are ready for meat.

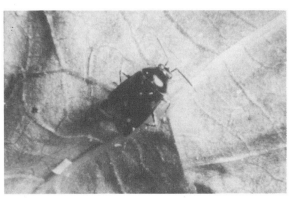
D. brevis stationed on a bean leaf.

D. brevis, like other predatory true bugs, attack their prey with a needle-sharp proboscis (see photo, next page for an example of a proboscis at work)

Photo by M. Herbut, courtesy of Applied Bio-nomics, Ltd.

through which pre-digestive enzymes are slowly exchanged for the bodily fluids of the prey. The proboscis is a sharp, straw-like mouthpart. It is certainly a weapon to be respected.

This is not a photo of D. brevis, but the nymph of another true bug, the brown stink bug, Euschistus servus, an excellent example shown here: proboscis erectus and a big meal.

The life-span of these predators is roughly 30 days in their immature stages, then around 21 days as adults. The conditions for optimum performance will be between 64-85°F in summer with a relative humidity of 30-60%; 73°F in winter to offset diapause (see Drawbacks). But these are *optimum* conditions, and not necessarily a prerequisite of successful implementation. Please note, however, cooler temperatures will hamper reproduction and development a certain degree.

Benefits— One-bug-does-all is not a bad trait of these predators. They are active searchers and flyers and can really kick some bug butt. Just looking at them and seeing their slight resemblance to the famed big-eyed bugs is enough to instill confidence.

D. brevis can be establish in a property if their needs are addressed. This *can* be economical. Simple amenities such as prey and pollen, upon which they can also feed, need to be available. Establishment is usually possible as long as the critical number of predators are introduced...

Drawbacks— The critical number, as mentioned above, pertains to the unusual directive which states that it is necessary to release 300 or more *D. brevis*, in *any* given area, in order for these predators to establish. We don't understand why this is so but, presumably, it is true.

Another drawback, but one hardly worth mentioning, is that these bugs, like the *Chrysoperla* spp., can deliver a painful little bite (to people and each other). This is not to scare you. It's insignificant compared the benefits. And never have we had negative client feedback regarding this slight drawback.

Photo by M. Cherim

Like *A. aphidimyza*, *D. brevis* undergo diapause (a quiescent state, hibernation) when the photoperiod is less than an estimated 10 hours (D) at 73°F. This is based on the recently established needs of *Orius insidiosus*, another true bug discussed in this manual (see thrips controls).

Availability is usually good. However, there are occasional production problems in the autumn. This irregularity is probably associated with the diapause reflex. These predators resent even small transitional changes to their schedule.

Scouting— Decimated pests and clean new growth are basically all the scout has to look for. However, with these bugs, decimated pests and clean new growth should both be abundant after a not-too-long period of time. Adults and nymphs might be found on flowers on occasion, but this is only obvious to the attentive scout as *D. brevis*, like many true bugs, are not gregarious and will hide upon your approach.

Advisories— To counteract the natural urge for these predators to undergo diapause in the winter, you must: 1) keep the temperature above approximately 73°F and, 2) provide supplemental lighting during the appropriate time of year. A 60 watt bulb for every 60 foot radius *may* do the trick. These predators are a little more difficult to fool than are *A. aphidimyza*. Bear in mind, also, most organisms, regardless of nature, will normally slow down a degree in the winter months.

A banker-crop of peppers may be beneficial. The predators enjoy existing on

This sweet pepper plant is showing severe aphid damage: notice the puckered, deformed leaves. D. brevis would love this.

Photo by M. Cherim

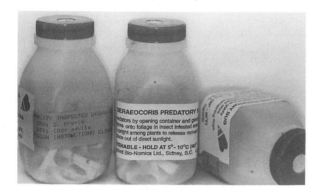

D. brevis units, the 100 count bottles.

pepper crops, as do aphids. Aphids can cause a tremendous amount of damage on peppers (see photo, previous page), *D. brevis* can help.

The use of these predatory bugs is still considered experimental — as is the case with all bio-controls — but is especially true with these bugs. One thing's for sure: these predators need *lots* of food.

Usages— *D. brevis* are useful in greenhouses, fields, interiorscapes, orchards and gardens. We recommend the successful implementation of these species in nearly every conceivable situation.

Rates—

```
PRVNT: N/A
CTRL-L: N/A
CTRL-M: 1-2 / YD. - BI-WKLY - 2-3X
CTRL-H: 2-4 / YD. - WKLY - 2-4X
MAINT: 1 / YD. - QTRLY - INDEF
GRDN: 40-60% of rates above.
ACRE+: 10-40% of rates above.
COMMENTS: Releases used for long-term
establishment should be of 300 or more bugs.
```

Pricing information—
The *D. brevis* true bug covered in this section can be obtained from The Green Spot by calling 603/942-8925. Detailed release instructions are provided with every order. Here is the current, industry-average pricing for 1998...

Item no. CDB1C ... 100 D. brevis adults ... 1 unit = $39.37

Item no. CDB1C-6 ... 100 D. brevis adults ... 6+ units = $35.43 ea.

Item no. CDB5C ... 500 D. brevis adults ... 1 unit = $173.25

Item no. CDB5C-4 ... 500 D. brevis adults ... 4-7 units = $155.92 ea.

Item no. CDB5C-8 ... 500 D. brevis adults ... 8+ units = $140.33 ea.

Photo by M. Cherim

The Convergent
Ladybird Beetle

Hippodamia convergens
(hip-oh-DAME-ee-ah CON-ver-genz)

Description— Ladybugs, as they called by the masses, are voracious predators. These orange/red, black-spotted beetles (see photo, below) are recognized the world over. They're the subject of a few songs, folklore tales, nursery rhymes and more. Nowadays, they're also an icon symbolic of bio-control and IPM. And the latter for good reason because, despite some of their unfavorable characteristics (see Drawbacks), they are effective preda-tors — or *can be* if used correctly.

Ladybug beetles definitely prefer to dine on **aphids**. They are considered an aphid predator. However, as most beetles are, *H. convergens* are very opportunistic and will eat pests other than aphids: **mites, insect eggs**, etc. We do NOT recommend using these beetles for other pests, though. We believe control of other pests cannot be obtained by typical ladybug releases. Coincidental cross-predation should be regarded as a bonus of the applica-tion and nothing more. We're aware of other claims of ladybugs being an end-all, cure-all predator, but it is regarded, at least by us, with scepticism, at best.

With aphids, though, the ladybugs have proven them-selves more than worthy. We have personally had phenomenal results with their use.

Adult ladybug devouring an aphid.

Photo by M. Badgley, courtesy of Buena Biosystems, Inc.

H. convergens are shipped as adults which have been field collection from their mass aggregation sites in the foothills of California's Sierra-Nevada mountains. Ladybugs are shipped to consumers directly from the collector/contractor out west (except for the two small sizes available through TGS).

Various quantities of the beetles (see Pricing information) are sacked in cotton/muslin sacks filled with excelsior/aspen strips to accommodate the predators during storage and travel.

Life-style— The large 8 mm. adult female beetles lay orange, football-shaped eggs usually on the upper-sides of the leaves of infested plants — up to 50 of them per day! The eggs hatch into black, alligator-like larvae with orange markings (see photo, below). These, too, are fierce predators, consuming 400 or so aphids.

The life-span of these predators is roughly 28 days in their immature stages, then around 11 months as adults. The conditions for optimum performance will be between 66-88°F with a relative humidity of 40% or greater. But these are *optimum* conditions, and not necessarily a prerequisite of successful implementation. Please note, however, cooler temperatures will hamper reproduction and development a certain degree.

The ladybug's hungry larval offspring is a serious threat to aphids

Photo by M. Badgley, courtesy of Buena Biosystems, Inc.

Benefits— Ladybugs can be very cost-effective. And at the recommended rates, they can be a very loud bang for the buck!

Another benefit is the IPM symbolism of ladybugs. They provide an excellent educational conduit between the grower and his or her customers. Most people will readily accept plants which are being treated with ladybugs. This can be an IPM medium upon which your customers' knowledge and acceptance of the Green Methods can flourish.

Most bio-control recommendations address the need for repeated releases. This is especially true with *H. convergens* (see Rates). This requirement can be easily satisfied with *H. convergens* without spending a great deal on the freight necessary to comply with the recommendations for these beetles. Why? They are refrigerator

A ladybug *ménage à trois*? "*Oui*, my loves, at the cucumber, tonight!"

storable (see Advisories). You can therefore buy in bulk and, if you do, you'll save on product *and* freight.

Drawbacks— First and foremost: they're flighty. If not released properly, ladybugs can disperse completely within 24 hours. Even when they are released properly, 90-95% will bolt anyway; it is expected, it is in their nature. We know of this phenomenon, and our release rates take this into consideration. It is the 5-10% which stick around that do the job for us (see Advisories for other info).

Another drawback, or more aptly labeled, an inconvenience, is the unavail-ability of these beetles around mid-May. Moreover, their springtime storage life, which is usually 2-3 months, is greatly reduced. This period marks the end of a ladybug generation.

Scouting— Decimated pests, clean new growth, adult and larval presence, and egg-clusters are indicators the scout should look for. The scout may find damage such as that shown in the photograph above. (However, after just a short time of using ladybugs, new growth will take off; the plant will start growing again. This is what happened to the pepper plant shown on page 39.

Photo by M. Cherim

Properly stored, ladubug beetles can last for months
in their storage bag with very little care.

It, as was several other plants, was badly infested. Ladybugs were released as per our instructions, and in two weeks, the plants were clean and new growth was plentiful. Ladybugs saved the day!)

Advisories— The refrigerator storage ability of *H. convergens* is more than just a convenience and cost-saver, it helps increase the effectiveness of these beetles. To do it properly try the following:

BEETLE STORAGE FACTS
1) Hold beetles between 35-45°F, better if cooler.

2) Store beetles in an older, not-so-frost-free fridge if possible.

3) Remove the beetles from the fridge once a month, mist the bags with water, allow to dry and return them to storage.

4) Hold beetles for no longer than 2-3 months.

5) Beetles purchased before May can be stored until mid-May to June first only; purchased after June first, the beetles can be stored full term. Many months under certain conditions.

Aside from watering the site before releasing and doing so in the evening, there are other things you can do to ensure the maximum number of beetles stick around. Flowering, pollen producing plants are a big plus (we carry a seed mix, Bioblend, with an appropriate variety of plants useful to ladybugs).

Pollen isn't the only thing these beetles will eat. They will also consume aphid honeydew. A product such as Biodiet™ (a honeydew substitute) may

Photo by D. Simser

help retain beetles.

Usages— Where can't they be used would a more appropriate question. We've seen the successful implementation of these species in just about every conceivable situation.

Rates—

```
PRVNT: N/A
CTRL-L: 4-6 / YD. - WKLY - 3-6X
CTRL-M: 6-9 / YD. - (every 4-5 days) - 4-8X
CTRL-H: 9-14 / YD. - (every 2-4 days) - 6-10X
MAINT: 3-5 / YD. - WKLY - INDEF
GRDN: 60-90% of rates above.
ACRE+: 15-50% of rates above.
COMMENTS: Releases larger that those recom-
mended can hamper program effectiveness.
```

Pricing information—
The *H. convergens* beetles covered in this section can be obtained from The Green Spot by calling 603/942-8925. Detailed release instructions are provided with every order. Here is the current, industry-average pricing for 1998...

Item no. CHCS1C ... 100 H. convergens w/ Biodiet ... 1 unit = $5.20

Item no. CHCS1C-6 ... 100 H. convergens w/ Biodiet ... 6+ units = $4.68 ea.

Item no. CHCS5C ... 500 H. convergens w/ Biodiet ... 1 unit =$10.80

Item no. CHCS5C-4 500 H. convergens w/ Biodiet ... 4+ units = $8.10 ea.

Item no. CHC45C ... ½ pt. H. convergens ... 1 unit = $12.60

Item no. CHC45C-6 ... ½ pt. H. convergens ... 6-11 units = $10.71 ea.

Item no. CHC45C-12 ... ½ pt. H. convergens ... 12+ units = $10.08 ea.

Item no. CHC9M ... 1 pt. H. convergens ... 1 unit = $16.64

Item no. CHC9M-4 ... 1 pt. H. convergens ... 4-7 units = $14.14 ea.

Item no. CHC9M-8 ... 1 pt. H. convergens ... 8+ units = $13.31 ea.

Item no. CHC18M ... 1 qt. H. convergens ... 1 unit = $24.64

Item no. CHC18M-4 ... 1 qt. H. convergens ... 4-5 units = $22.17 ea.

Item no. CHC18M-6 ... 1 qt. H. convergens ... 6+ units = $19.71 ea.

Item no. CHC36M ... ½ gal. H. convergens ... 1 unit = $45.76

Item no. CHC36M-2 ... ½ gal. H. convergens ... 2-3 units = $41.18 ea.

Item no. CHC36M-4 ... ½ gal. H. convergens ... 4+ units = $38.89 ea.

Item no. CHC72M ... 1 gal. H. convergens adults ... 1 unit = $63.10

Item no. CHC72M-2 ... 1 gal. H. convergens adults ... 2-3 units = $56.79 ea.

Item no. CHC72M-4 ... 1 gal. H. convergens adults ... 4+ units = $53.63 ea.

NOTE: Another beetle suitable for aphid control is the multi-colored Asian ladybeetle, Harmonia axyridis (see scale insect controls).

caterpillars

Order: Lepidoptera

Caterpillars, loopers, worms, etc., are the larvae of moths and butterflies. Above, only the order was given because this broad pest group is comprised of the members of a multitude of families. There are over 110,000 members, (ranked number two worldwide —beetles [Coleoptera] are number one with 300,000+ species).

Caterpillars can be very frustrating pests to have. They'll usually make themselves known by the damage; and the damage is typically noticed when the situation is pretty much out of hand. This is a classic example of when record keeping really comes in hand: last year's fiasco can become this year's advance warning.

Starting a caterpillar control program doesn't mean you have to wait until the following year, hoping to catch the newly-hatched caterpillars early. You can start a whole lot earlier. Today's caterpillars are tomorrow's pupae [chrysalis], (though the time of year will vary from species to species.)

Some caterpillars pupate in the soil, and in such cases, they may be able to be controlled there: using a soil-dwelling pest control (see the applicable section), perhaps. Some caterpillars lay eggs in the late summer, they, too, may be able to controlled there: with an egg parasite or smothering oil, perhaps.

In some cases, though, the only thing you can effectively do is to treat the caterpillars themselves. Bt, neem, Beauveria bassiana, etc. are all tools at the grower's disposal. (Learn more about these products in this manual.)

The adult European Gypsy moth
Lymantria =Porthetria dispar

Illustration by M. Cherim

Keep in mind, to control caterpillars effectively you'll have to work at it. Be prepared by knowing the following things: 1) be familiar with the pest species and its life-cycle, timing, etc.; 2) know how you're going fight back (especially when) and; scout, scout, scout, don't miss your window of opportunity, which, with some species, is very narrow.

Here's an example of a caterpillar control

program for the European Gypsy moth (see illustration, facing page): the non-flying female moths lay eggs in the latter part of the summer (release parasites, smother eggs in horticultural oil, use dormant oil in winter, destroy egg sacks by hand); the remaining eggs hatch in the spring (use egg parasites before eggs hatch, buy or attract predators which may impact the eggs.); the eggs hatch into tiny caterpillars which feed on the leaves of the tree upon which the eggs were deposited (band the trees with a sticky compound to prevent the caterpillars' movement, and to trap some of them); the caterpillars grow in size, feeding all the while, until they "balloon" or hang from a silken thread until the wind takes them to their new home); they pupate, hatch into adults, mate and start laying eggs (this is where we came in).

A parasitic braconid (brah-KOH-nid) wasp ovipositing, or laying eggs in a tomato hornworm.

Usually the first sign of caterpillars will be the absence of massive amounts of greenery. These pests can really cause a lot of damage to trees, bushes, etc. Another indicator, besides that half the tree is gone, will be fecal matter found below the feeding site. An example would be the tomato hornworm: you can actually hear them chewing; they are about the size of

The baby wasps will form cocoons on the back of the caterpillar when they pupate, these are NOT baby caterpillars.

a freight train; beautifully colored, but only when you can find them on the plant; then you notice the tons of very dark green droppings on some lower leaves — look up, there it is. Now check out the caterpillar to see if he/she is the victim of parasitic wasps which deposit eggs in the hornworm. (See photos, above). If so, remove the worm from the plant, but do not kill it — let the parasitoids hatch.

Greenhouse and interiorscape infestations of caterpillars happen infrequently. However, if the situation arises, just follow the guidelines detailed previously. For more detailed information and recommendations concerning caterpillars, loopers, worms, etc., and ways to control them, you are invited to call The Green Spot, Ltd. at 603/942-8925. 🞖   Both photos by M. Cherim

caterpillar controls

The Lepidopteran

Egg Parasitoids

Trichogramma spp.
minutum and pretiosum

(tree-KOH-grah-ma MY-nu-tum, pret-ee-OH-sum)

Description— These 0.9 mm. mini-wasps are very popular and very effective *Lepidopteran* egg parasitoids. *Lepidoptera* (lep-ih-DOP-ter-rah) is the name of the order to which moths and butterflies belong. The eggs of over 150 species assigned as members of this order can be parasitized by the *Trichogramma* spp.

T. minutum, which are shipped as pre-parasitized, ready-to-hatch grain moth eggs (*Sitotroga cerealella*) adhered to a card which is perforated into 30 squares (as packed, see photo to the right), are the product of choice when in an orchard or tall crop setting. In *this* case, "tall crop" means one which is over the 8-10 foot range.

T. pretiosum, which are shipped in the same style and manner as *T. minutum*, above, are the product of choice when in a field, greenhouse or short crop setting. In *this* case, "short crop" means one which is under the 8-10 foot range.

Application rate protocols for both of the *Trichogramma* spp. described above remain unchanged, the height of the crop in the case of these mini-wasps determines the proper species to choose. (For more info, see Rates.)

Some popular hosts of these mini-wasp include the eggs of: the **Gypsy moth** (shown above, exposed eggs only); **codling moth (*Cydia pomonella*)**; **diamondback moth (*Plutella xylostella*)**; **Oriental fruit moth (*Graphiolitha molesta*)**; **tomato pinworms (*Keiferia lycopersicella*)**; **cabbage loopers (*Trichoplusia ni*)**; **imported cabbage worms (*Pieris rapae*)**; **tent caterpillars (*Malacosoma* spp.)**; and many, many more, even the grossly damaging **tobacco/tomato hornworms (*Manduca* spp.)**.

Life-style— The *Trichogramma* spp., as parasitoids, work by laying eggs in the eggs of many *Lepidopteran* spp. as mentioned under Description, above. The adult wasps can lay up to 300 eggs each, parasitizing an equivalent number of soon-to-be destructive caterpillars, loopers and/or worms (see photo, right).

Female Trichogramma sp. wasp sampling an egg for its ability to host its young.

The wasps' larvae which hatch from the eggs, attack the moths' eggs' embryos from within (*endo*parasitism). Instead of pests hatching out, more mini-wasps hatch out instead.

The life-span of these parasitoids is roughly 7 days in their immature stages, then up to 10 days as adults. With such an abbreviated life-cycle, these parasitoids can sometimes build up to 30 generations per year (many of which overwinter). The conditions for optimum performance will be between 70-85°F with a relative humidity of around 60%. But these are *optimum* conditions, and not necessarily a prerequisite of successful implementation. Please note, however, cooler temperatures will hamper reproduction and development a certain degree.

Benefits— These wasps are cheap, easy and effective. The cards that these parasitoids are supplied on (see photo, next page), because they are perforated into squares, as previously mentioned, make even distribution a sure thing. There's only one problem, see next...

Drawbacks— Ants. They *love* the eggs and will rob them from the squares. Therefore creativity is sometimes necessary: stapling the squares to leaves, attaching them to a tree trunk and surrounding the small area with a sticky barrier product may help.

Another option is to pre-hatch the eggs. This is covered in the instructions which accompany shipments.

Photo by M. Badgley, courtesy of Buena Biosystems, Inc.

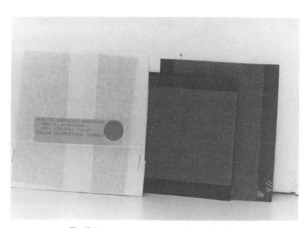

The Trichogramma spp. are supplied as pupae in grain moth eggs adhered to a perforated card.

Another drawback is their difficulty to scout (see Scouting).

And, as mentioned before, the *Trichogramma* spp. have a wide host range. This means they can parasitize a great number of eggs belonging to a great number of species. These wasps are indiscriminate killers. And as much as this is good and convenient to the grower, it can be hazardous to nontarget organisms. Caution and awareness should be exercised when employing these awesome parasitic mini-wasps (see Advisories for more details).

Last drawback: you must really know the pest you're dealing with. Timing is critical in many cases. The eggs have to be available, and one sure way to guarantee this is to know when they are being deposited. The *Trichogramma* spp. are only a preventive weapon, they can't do them in once the larvae hatch.

Scouting— Your hands are tied. Seeing these little wonders at work is normally out of the question. They're just too small and quick. If you can find the moth eggs, you may be able to determine if they have exit holes from the wasps. The only other indicator is the reduction of the numbers of caterpillars upon their hatch, and [less] damage noticed shortly thereafter.

Some caterpillar pests are more difficult to control than others. This can be a drawback, especially if the tough-to-kill pest is a serious pest. The Gypsy moth is a good example of this. With no significant numbers of natural enemies (because this pest was *introduced* to the U.S. from abroad), *Trichogramma* spp. may have a difficulty making an impact if used by themselves. Therefore the implementation of multiple IPM resources is necessary. Unfortunately, this costs time and money.

Battling ants, which are a real threat to bio-control when it comes to these

Photo by M. Cherim

wasps, is sometimes very difficult. It adds complexity to the would-be simple implementation of these parasitoids.

Advisories— If you are aware of an endangered species of moth or butterfly at your release site, the release should be curtailed, or at least minimized. These wasps are not long-range flyers, though, your release doesn't have to be too far from the wildflower meadow. Use common-sense, that's all.

Now speaking of wildflower meadows, this points us towards an interesting topic: trap-crops. But not in the conventional sense. Normally trap-crops will harbor the beneficials *and* pests, but in the case of *these* pests, a trap-crop probably won't lure them. Moth and butterfly larvae seem to be such specific feeders. You will, however, by using a close-range trap-crop [banker-crop], lure other beneficials. We see caterpillars being dragged off by wasps, birds and bugs all the time. Don't spray pesticides (*Bacillus thuringiensis*, *Bt*, etc. not included), and lure the good guys. These two steps can do an awful lot and may lessen the necessity of purchased-parasitoid use.

Usages— Greenhouses and interiorscapes have employed these wasps. Mainly, however, these parasitoids are used in field and row crops, orchards and gardens.

Rates—

```
PRVNT: 80-160 / YD. - WKLY - NEED
CTRL-L: N/A
CTRL-M: N/A
CTRL-H: N/A
MAINT: 20-30 / YD. - MTHLY - INDEF
GRDN: 55-85% of rates above.
ACRE+: 25-65% of rates above.
COMMENTS: Timing is critical. The window of
opportunity for prevention must be observed.
```

Pricing information— The *Trichogramma* spp. wasps covered in this section can be obtained from The Green Spot by calling 603/942-8925. Detailed release instructions are provided with every order. Here is the current, industry-average pricing for 1998...

Item no. CTM120M ... 120k (card) T. minutum ... 1 unit = $18.72

Item no. CTM120M-4 ... 120k (card) T. minutum ... 4-5 units = $17.78 ea.

Item no. CTM120M-6 ... 120k (card) T. minutum ... 6-11 units = $16.94 ea.

Item no. CTM120M-12 ... 120k (card) T. minutum ... 12+ units = $15.58 ea.

Item no. CTP120M ... 120k (card) T. pretiosum ... 1 unit = $18.72

Item no. CTP120M-4 ... 120k (card) T. pretiosum ... 4-5 units = $17.78 ea.

Item no. CTP120M-6 ... 120k (card) T. pretiosum ... 6-11 units = $16.94 ea.

Item no. CTP120M-12 ... 120k (card) T. pretiosum ... 12+ units = $15.58 ea.

leafminers

Order: Diptera   Family: Agromyzidae

So the leaves of your plants have tunnels in them? Looks like some mesophyll (the cells between the outer leaf surfaces) is missing? Sounds like leafminers. But are they a big problem? Well, some say no. It is true that natural enemies to these pests can effectively control them — assuming previous spray applications have not killed them off. Woody ornamental plants, especially those slated for resale, tolerate few leafminers. Not that they're a detriment to the health of the plant (unless the infestation is serious), but a lot folks will notice and not accept the damage. However, if the plants are relocated to a natural setting with a heathy ecosystem, good soil, light, air, water and the such, the leafminers should not be an unconquerable problem. But this is outdoors, too.

A greenhouse or indoor infestation of leafminers, especially if the systems are "closed," can be an entirely different matter. A crop of tomatoes could be jeopardized by leafminers. The same holds true for a fairly short-season ornamental crop grown outdoors such as chrysanthemums. In these situations, control is more critical.

Break out the big guns? Not yet. Neem is just one of many products available and labeled for leafminers in a multitude of crops. There are also two common parasitoids that are quite effective (covered next). Since many leafminers pupate in the soil, soil-dwelling pest controls (see the applicable section) can also effectively contribute to the control of these little flies.

Scout for the flies, but don't expect to see anything striking. More often than not it's the damage which is first discovered, then the larvae — snug in their ever-lengthening mines, then the adults are seen (unless the scout is really aware of what she or he is looking at on their yellow sticky traps.

For more detailed information and recommendations concerning leafminers and ways to effectively control them, you are invited to call The Green Spot, Ltd. at 603/942-8925.

leafminer controls

The Cool-Weather

Leafminer Parasitoid

Dacnusa siberica

(DAK-nuus-ah sy-BEER-ih-kah)

Description— These 3 mm. mini-wasps are best used for preventing the establishment of several leafminer species. They can also tackle minor infestations. And, if established, they can adequately protect a crop throughout the [off-] season.

D. siberica, which are shipped as pre-hatched, pre-fed adults, are the product of choice when leafminers are first seen, or soon expected to be seen, in a cooler area. They do best in cooler conditions, as the species name implies: *siberica*, brrr!

Some of the several species which can be controlled with these parasitoids include the **Florida, chrysanthemum or serpentine leafminer (***Liriomyza trifolii***)**, the **tomato leafminer (***L. bryoniae***)** and **many others of economic importance**.

Life-style— These parasitoids, work by laying eggs directly into the larvae of leafminers. And they can lay up to 72 eggs, killing a like number of the larvae. They do this while the larvae are working in their mines or tunnels. These mines are created as the larvae consume the leaf's mesophyll (the cells which make the center of the leaf sandwich). The wasps' larvae which hatch from the eggs, slowly weaken and kill the leafminer larvae from within (*endo*parasitism). New parasitoids hatch out of the mines instead the fattened, full-term leafminer larvae, which usually drop to the ground to pupate (see Advisories).

The life-span of these parasitoids is roughly 2 weeks in their immature stages, then 1 week as adults. The conditions for optimum performance will be between 55-75°F with a relative humidity of around 70%. But these are *optimum* conditions, and not necessarily a prerequisite of successful implementation. Please note, however, *extremely* cool, or *warm*, temperatures will hamper reproduction and development a certain degree.

Benefits— The life-cycle of these parasitoids is considerably shorter than that of their hosts. They can, therefore, overwhelm the pests in a matter of time with few, if any, other considerations. Patience is it. Under most conditions these mini-wasps will prosper and come out on top.

As described under Life-style, the tunneling which occurs is the damage of importance. It creates a visual eyesore. These parasitoids will help lessen the immediate damage, but they can't reverse the damage [that's why these wasps are a good late-winter/early-spring preventive]. The main benefit, however, if used when leafminers are present, is to diminish the number of future pest generations.

If bad enough, the mines can become more than just a visual problem, they can hinder the plants' photosynthesis. They can kill a plant! Therefore, ornamental crops are not the only applicable crops to use these on. Vegetables, too, are a candidate.

Dacnusa siberica are superb preventive agents, thus offering growers a potential money-saving tool. Additionally, they can establish themselves in nearly any situation. Unless things get too hot. (For warmer situations, see *Diglyphus isaea*, next) Once established, growers might be able to reduce the size of releases made due to the presence of on-site wasps: another money-saver.

D. siberica, being shipped as adults, offer the benefit of fast oviposition or egg-laying. And they can do this is in temperatures so low that many other bio-control agents would stop in their tracks.

Drawbacks— Despite the pages and pages of information we're offering about this species of good critter, we know relatively little. Our exposure to practical experiences is very limited. Try it with all of Europe's widespread successes confidently behind you, but do so with the cautiousness and curiousness of a researcher.

These parasitoids are produced in Europe. This, among other things, adds to their expense. This drawback may disadvantage smaller growers (because a little bit goes a long way). We are told that they are well worth the money, though.

These parasitoids are incapable of handling large infestations in an acceptable period of time; not cost-effectively, anyway. There can sometimes be availability problems, so plan well ahead.

Scouting— Reduction of pest numbers is a sure sign of success. Or, if used preventively, a pest no-show.

The scout can sometimes also find adults on affected leaves. They'll be seen tracing the mines with their antennae; this is how they find the location of their hosts. Exit holes may also be seen, but this extremely difficult to properly determine.

Advisories— Yellow sticky traps *may* present a problem with these wasps. We are unsure of this and ask that you exercise cautious observation when the traps are in use. If parasitoid numbers on the traps get too high, remove the traps or use them for only two or three days per week.

As explained under Life-style, the leafminer larvae drop to the ground to pupate. And now for some good news: they can be parasitized by our beneficial nematodes (see Soil Pest Controls).

Usages— Greenhouses, fields, interiorscapes, orchards and gardens. These wasps are reportedly successful at nearly any site, 'cept a hot one!

Rates—

PRVNT: 1-3 / YD. - BI-WKLY - NEED
CTRL-L: 2-5 / YD. - BI-WKLY - 2-3X
CTRL-M: N/A
CTRL-H: N/A
MAINT: 2-4 / YD. - MTHLY - INDEF
GRDN: 40-60% of rates above.
ACRE+: 15-45% of rates above.
COMMENTS: Early season introductions will work the best. Use with parasitic nematodes.

Pricing information— The *D. siberica* wasps covered in this section can be obtained from The Green Spot by calling 603/942-8925. Detailed release instructions are provided with every order. Here is the current, industry-average pricing for 1998...

Item no. CDS250 ... 250 D. siberica adults ... 1 unit = $29.40

Item no. CDS250-4 ... 250 D. siberica adults ... 4-7 units = $27.93 ea.

Item no. CDS250-8 ... 250 D. siberica adults ... 8+ units = $25.14 ea.

The Warm-Weather
Leafminer Parasitoid

Diglyphus isaea
(di-GLIF-us i-zay-ah)

Description— These 2 mm. mini-wasps, like *D. siberica*, are best used for preventing the establishment of several leafminer species. They can also tackle minor infestations. And, if established, they can adequately protect a crop throughout the season.

D. isaea, which are shipped as pre-hatched, pre-fed adults (see photo, facing page), are the product of choice when leafminers are first seen, or soon expected to be seen, in a warmer area. They do best in warmer conditions.

Some of the several species which can be controlled with these parasitoids include the ones controlled by the preceding bio-control agent. They **can control 18 known species from four different genera**, not just the genus *Liriomyza*.

Life-style— These parasitoids, work by "stinging" (oviposition) the larvae of leafminers while they work in their mesophyll mines (as described under *D. siberica*) with their ovipositors (egg-laying apparatus). The stinging conveys a paralyzing toxin to the pests. The toxin works right away — the larvae stop feeding. But these mini-wasps don't stop there. They will also feed on the hosts (host-feeding), as do many *Chalcid* (KAL-cid) wasps (a wasp family). They will then lay 1-5 eggs next to the paralyzed larvae — up to 50 eggs in their life. The wasps' larvae which hatch from the eggs, then consume the leafminer larvae from the outside-in (*ecto*parasitism). The parasitoids' larvae then construct pillars of fecal matter around the remains of the much deteriorated pests. These are thought to protect the beneficial larvae from leaf drying, etc., while they undergo pupation.

The life-span of these parasitoids is roughly 2 weeks in their immature stages, then 3 weeks as adults. The conditions for optimum performance will be between 75-90°F with a relative humidity of around 80%. But these are *optimum* conditions, and not necessarily a prerequisite of successful implementation. Please note, however, cooler temperatures will hamper reproduction and development a certain degree.

Benefits— The benefits are very much the same as those of *Dacnusa siberica* (see previous for details), with exception to the time of year for best use (spring and summer) and their temperature preferences.

D. isaea, also being shipped as adults, offer the benefit of fast oviposition or egg-laying.

D. isaea "feeling" the leaf for a good oviposition site.

Drawbacks— The same drawbacks which plague the previously discussed parasitoid, hamper the use of this one too. This includes the price issue. But, as is the case with *D. siberica*, a little bit goes a long way.

Also, there can sometimes availability problems, so plan well ahead.

Scouting— Reduction of pest numbers is a sure sign of success. Or, if used preventively, a pest no-show. The scout can sometimes see the loose fecal matter on the leaves (cause by the paralyzation process), or the organized fecal matter in the form of pillars used for the pupal tombs.

Advisories— Yellow sticky traps *may* present a problem with these wasps. We are unsure of this and ask that you exercise cautious observation when the traps are in use. If parasitoid numbers on the traps get too high, remove the traps or use them for only two or three days per week.

Usages— Greenhouses, fields, interiorscapes, orchards and gardens. These wasps are reportedly successful at nearly any site, 'cept a cool one!

Rates—

```
PRVNT: 1-2 / YD. - TRI-WKLY - NEED
CTRL-L: 2-4 / YD. - BI-WKLY - 2-3X
CTRL-M: N/A
CTRL-H: N/A
MAINT: 2-3 / YD. - MTHLY - INDEF
GRDN: 45-65% of rates above.
ACRE+: 20-40% of rates above.
COMMENTS: Late spring introductions will
work the best. Use with parasitic nematodes.
```

Pricing information—
The *D. isaea* wasps covered in this section can be obtained from The Green Spot by calling 603/942-8925. Detailed release instructions are provided with every order. Here is the current, industry-average pricing for 1998...

Item no. CDI250 ... 250 D. isaea adults ... 1 unit = **$61.42**

Item no. CDI250-4 ... 250 D. isaea adults ... 4-7 units = **$56.81 ea.**

Item no. CDI250-8 ... 250 D. isaea adults ... 8 + units = **$52.55 ea.**

Photo by M. Badgley, courtesy of Buena Biosystems, Inc.

scale insects

Order: Hemiptera   Sub Order: Homoptera
Families: Coccidae, Diaspididae & Pseudococcidae

Just so there is no immediate confusion, scale insects, as a group, contain three distinct members (families): mealybugs (Pseudococcidae), armored (Diaspididae) and soft (Coccidae) scale insects. Instead of addressing mealybugs under their own section, we felt it was more accurate to list them here.

Scales, including mealybugs, are some of worst pests a grower can encounter. The price of control can be very high — even when using chemical and conventional means. The main problem with scales is they have built-in protection. The immature stages can be reasonably dealt with, but the adults are tough as nails; they all have some sort of armor. The armored and soft scales both have a tough, chemical- and predator-resistant covering. Mealybugs are covered in a waxy covering (see photo, below). It's not a tough material like on the other scales, but it is pretty impervious to many sprays.

Scales suck on fluid from the plant and, like aphids, can cause an unbelievable

Citrus mealybugs (Planococcus citri) are covered
and protected by a white, waxy substance.

Photo by M. Herbut, courtesy of Applied Bio-nomics, Ltd.

amount of damage. In some species of mealybug, chemical in the saliva can cause the leaves to curl around it, thus protecting it further.

Soft scales and mealybugs also excrete honeydew, a sticky, sugary by-product of their feeding. This may attract ants which will further protect the pests from natural or introduced bio-control agents. (See Aphids, pages 24 & 25.)

Typical infestations begin by a single female waiting for a winged adult male to land on the stressed-out plant she's calling home. Eggs are laid; they hatch, the offspring grow and reproduce themselves, yielding more young. As feeding and reproducing continues, the honeydew gets thick and the plant starts to signs of serious damage. This is, unfortunately, how most infestations are noticed. And controlling them at this point can be hard.

Scouting is extremely important. Interiorscapers, make sure your techs know how to look, and what to look for. Catching scale insects and mealybugs early is fundamentally important — it can save the plants.

Seeing flying things you haven't seen before? Got bumps on your branches or fruit? Sticky stuff and ants? White, cotton-like material? Small white lines on the trunk? These are all signs of scale or "mealy" infestation.

The Green Spot may be able to assist you in the identification of your scale insects (incl. mealybugs). Call us if we're needed.

For more detailed information and recommendations concerning scale insects and ways to effectively control them, you are invited to call The Green Spot, Ltd. at 603/942-8925. ✄

scale controls

The Golden Chalcid

Armored-Scale

Parasitoid

Aphytis melinus
(ay-FIH-tiss mel-LINE-uss)

Description— These yellow-gold, 1.2 mm. mini-wasps are best used to tackle and maintain minor to heavy infestations of several armored scale species. And, if established, they can adequately protect a planting for a season or, perhaps, more in certain situations.

A. melinus, which are shipped as pre-hatched adults, are the product of choice when hard scales that can host the Golden Chalcid are present. These wasps are shipped with Biodiet™ on the lid of their jar, but it is provided merely as a moisture source. Armored scale insects do not produce honeydew and, therefore, the mini-wasps are not accustomed to it.

Some of the species which can be controlled with these parasitoids include the **California red** (their favorite) and **yellow scale insects (*Aonidiella aurantii and A. citrina*, respectively); the San Jose scale (*Quadraspidiotus perniciosus*); the ivy or oleander scale (*Aspidiotus nerii*); the oystershell scale (*Lepidosaphes ulmi*)** and **other economically important species including,** as we discovered in 1997, the **magnolia white scale (*Pseudaulacaspis cockerelli*).**

Life-style— These parasitoids, which find their hosts with their antennae or feelers (see photo, facing page), work by laying eggs underneath female scales which have loosened their position on the leaf to facilitate reproduction. They will lay up to 25 eggs, killing 5 to 25 pests as they do. They will feed on many more. The wasps' larvae which hatch from the eggs, then methodically eat their way through the scales from the underside out through the back (*ecto*parasitism). New parasitoid(s) hatch out of the empty scale cadavers after they pupate. And they're off again!

The life-span of these parasitoids is 18 days in their immature stages, then not quite a month as adults. The conditions for optimum performance will be between 76-85°F with relative humidity between 40-50%. But these are *optimum* conditions, and not necessarily a prerequisite of successful implementation. Please note, however, cooler temperatures will hamper reproduction and development a certain degree.

The antennae are used to locate scale insects suitable to host the wasp's offspring. The antennae's hairs are extremely sensitive tools.

Benefits— They're quite inexpensive and, if you've got the right host and conditions, they're an incredible buy!

A. melinus are superb curative agents, thus offering growers a potential money-saving tool (they're a lot more economical than pesticides). Additionally, they can establish themselves in nearly any situation. Once established, growers might be able to reduce the size of, or curtail, future releases due to the presence of on-site wasps: another money-saver.

Lastly, these parasitoids, being shipped as adults, offer the benefit of fast oviposition or egg-laying, plus a lot of host-feeding. Yup, these guys too!

Drawbacks— These mini-wasps have a low fecundity (egg-laying) rate, are short-lived and they won't overwinter in cold climates. Nevertheless, they do a good job, are supplied in a large quantities and are affordable. This leaves us with only one more possible drawback...

These wasps can be difficult to scout (see next).

Scouting— Reduction of pest numbers (empty scales and lack of crawlers) is a sure sign of success.

The scout can also usually find scale insect cadavers with an exit hole in them. This is a very positive sign.

Trying to spot the wasps themselves, say, two weeks after a release, would be an exercise in futility.

S.E.M. photo by N. Cherim

A. melinus "sizing" up a host by way of a tapping dance.

Something learned in 1997 was found to be of considerable value, especially since *A. melinus* is the only widely available parasitoid of armored scale. It works like this: Say you don't have a clue to what species of armored scale you have, and don't want to go through the trouble to have it identified, or you do know what it is, but it's not on the known-host list for *A. melinus*. Well, get some of these wasps anyway (the cost is minimum), and carefully observe them when you place them in very close proximity to the pests. If they investigate then do a little tapping dance on the scales, they're sizing them up for the ability to host their young (see photo, above). This is great news for you. If not, and the parasitoids simply walk away, ignoring the scales, then it might be time to break out the hort-oil.

Advisories— If armored and soft scales are present together (soft scales being a honeydew producer), ants may be a problem with this parasitoid. They don't seem to mind the sticky stuff themselves, though. Use barrier products or boric acid products to control the ants, if they are a problem.

Usages— Greenhouses and interiorscapes are the settings where these parasitoids are most often employed. However, in the right environment, these wasps can be used with great success outdoors. After all, they *are* reared especially for the California citrus industry.

Rates—

```
PRVNT: N/A
CTRL-L: 16-24 / YD. - TRI-WKLY - 3-4X
CTRL-M: 18-36 / YD. - BI-WKLY - 3-4X
CTRL-H: 24-48 / YD. - WKLY - 3-5X
MAINT: 5-9 / YD. - MTHLY - INDEF
GRDN: 40-65% of rates above.
ACRE+: 15-45% of rates above.
COMMENTS: Outdoors, time your first
inoculations with the flight of scale males.
```

Pricing information—
The *A. melinus* wasps covered in this section can be obtained from The Green Spot by calling 603/942-8925. Detailed release instructions are provided with every order. Here is the current, industry-average pricing for 1998...

Item no. CAM10M ... 10,000 A. melinus adults ... 1 unit = $24.64

Item no. CAM10M-8 ...10,000 A. melinus adults ... 8+ units = $22.67 ea.

Item no. CAM30M ... 30,000 A. melinus adults ... 1 unit = $56.16

Item no. CAM30M-6 ...30,000 A. melinus adults ... 6+ units = $51.67 ea.

The Mealybug Destroyer
a.k.a. "Crypts"

Cryptolaemus montrouzieri
(krip-toh-LAY-muss mon-TROH-zuur-ee)

Description— Like *Hippodamia convergens*, the popular aphid predators, *C. montrouzieri* are ladybug beetles. Unlike their relative, though, these beetles are not wild-collected, do not aggregate and do not disappear just after release.

Crypts, with their shiny black body and dull-orange head and thorax, definitely prefer to dine on **mealybugs ('specially young 'uns)**. They can clean up large populations. However, as most beetles are, *C. montrouzieri* are very opportunistic and will eat pests other than mealybugs: **other scale insects (their crawlers or immature forms), insect eggs, etc.** We do NOT recommend using these beetles for other pests, though. We believe control of other pests might not be obtainable with typical Crypts releases. Coincidental cross-predation should be regarded as a bonus of the application and nothing more.

With mealybugs, though, these beetles have proven themselves more than worthy. Our customers have reported phenomenal results with their use. Especially our interiorscaper and botanical garden customers.

C. montrouzieri are shipped as pre-fed, pre-mated, insectary-reared (for the citrus industry) adults (see photo, right). *Some* popular prey of these beetles include: the **citrus mealybug (*Planococcus citri*); the comstock mealybug**

C. montrouzieri beetle feasting on mealybug eggs.

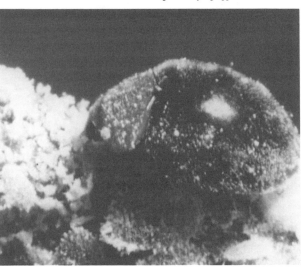

Both photos, facing page and above, by M. Badgley, courtesy of Buena Biosystems, Inc.

The citrus mealybug, and others, produce a
cottony mass in which to lay their eggs.

(*Pseudococcus comstocki*); the obscure mealybug (*Pseudococcus obscurus*); the solenopsis mealybug (*Phenacoccus solenopsis*); the Mexican mealybug (*Phenacoccus gossypii*) and many other related species, even the long-tailed mealy-bug (*Pseudococcus longispinus*) can be consumed with greedy abandon, but only if it is present with another species which produce cottony egg-masses (see Advisories).

Life-style— The large 5 mm. Australian, adult female beetles lay their eggs in the cottony egg-masses produced by the pests (see photo, above) — one egg per mass, usually, and up to 10 of them per day, for up to 50 days! The

This Crypt's larva is a wolf in sheep's clothing.

eggs hatch into white, shaggy-coated larvae (see photos, left and facing page) which, to the inexperienced, look like mealybugs (a wolf in sheep's clothing). These, too, are fierce predators, growing up to 1 cm. long and consuming 250 or so small mealybugs and their eggs

Photo, top, by L. Gilkeson, courtesy of Applied Bio-nomics, Ltd.
Both photos, bottom and facing page, by D. Simser

(they'll always eat the young-est, most tender morsels first).

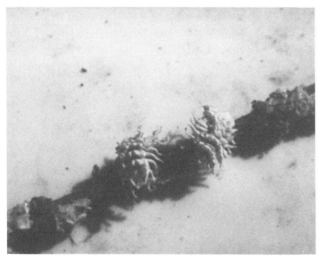

One major difference between the Crypts' larvae and the citrus mealybug: the wax, filamental hairs of the C. montrouzieri larva are long and "wild."

The life-span of these predators is roughly 3½ weeks in their immature stages, then around 1½ months as adults. The conditions for optimum performance will be between 64-91°F with a relative humidity of 70% or greater. But these are *optimum* conditions, and not necessarily a prerequisite of successful implementation. Please note, however, cooler temperatures will hamper reproduction and development a certain degree.

Benefits— Considering how good the performance of these predatory beetles is, they are a cost-effective solution to your mealybug problems.

These beetles are usually easy to scout (see Scouting).

Drawbacks— They won't fly at temperatures below 56°F. They will still work, they don't undergo diapause (a quiescent state, hibernation), they will just do so more slowly and inefficiently.

Another drawback is they seem to be a temporary fix in interiorscapes, providing outstanding results, but for only 8-26 weeks. The adult beetles leave when they are *nearly* done, and their young must die.

But, thinking about it, isn't all pest control just temporary? Consequentially, and because our customers say that the Crypts are still the most economical and easiest control, we have to stop and wonder if there are *any* significant drawbacks concerning the use of this species. Perhaps not.

Perhaps so! *C. montrouzieri*, at least in most of 1997, were very difficult to obtain in significant numbers, i.e., 12,800 were needed one week, but only 300 were received. This, however, should not stay in the Drawbacks section for too long; Crypts are much too important of a bio-control agent to allow this to continue. Demand is very high.

Scouting— Adult beetle presence (see photo, below), larval presence, reduction of pest numbers, "exploded" mealybug egg-masses. These are all signs that Crypts are hard at work.

Talk about presence, here's 500 or so beetles.

Now, having written that, a lot of interior-scapers are probably wondering if this easy-to-spot predator will be noticed by their clients. The answer is *probably* not. We've never heard of a client complaint about Crypts in their planting(s) — so far. We strongly suggest that interiorscape contractors *do* consult with their clients before putting in the predators.

Advisories— Crypts need to lay their eggs in the cottony egg-masses of their prey (as discussed under Description and Life-style). Since long-tailed mealybugs don't lay eggs, but rather give live-birth, as aphids do, they provide no cottony masses in which the beetles can lay their eggs. This is not a problem if your site has more than one species (assuming the other species *does* produce the egg-masses). Crypts will *eat* immature long-tailed mealybugs with great joy, they just can't *reproduce* on them.

The problem of a long-tailed-mealybug-only infestation can be overcome a couple of different ways:

> 1) You can place little bits of synthetic quilt batting (available from craft stores) amongst the long-tailed mealybug populace. (Avoid cotton balls, they may contain pesticide residues.) We're not completely sure how well this works, but some of our customers have tried it, experimentally, and reported satisfactory results. Also, try placing white 3" x 5" index cards, halved, by groupings of the pests to

Both photos, above and facing page, by M. Cherim

lure them in.

2) Employ the *Chrysoperla* species (green lacewings, see Aphid Controls) at the site. We've been told on numerous occasions that they do an excellent job. Use only the eggs or the larvae, however. Preferably the larvae. We don't think the adults will do too well.

Aside from misting the site with water before releasing and doing so in the evening (sometimes not necessary in interiors), there are other things you can do to ensure the maximum number of beetles stick around. Flowering, pollen producing plants are a big plus to Crypts).

Pollen isn't the only thing these beetles will eat. They will also consume mealybug honeydew; they produce a lot. A product such as Biodiet™ (a honeydew substitute) may help encourage the beetles. In fact, no one other of our organisms like our product more. We always ship Crypts, as we do with several of our organisms, with an in-flight snack of the stuff. And they *never* complain about the airline's cuisine.

Ants, if present, should be controlled. They will defend mealybugs from predators and parasites to protect their honeydew/excrement food, blah! Use barrier products or boric acid products to control the ants.

In 1996 we were contacted by a popular professional interiorscape writer and speaker, Linnaea Newman, with a new and ingenious way to release Crypts in interiorscape trees: Find a long bamboo or wooden pole with an approximate diameter of three-sixteenths to one-quarter-inch; insert one end of the pole through the pre-manufactured hole in our jars' lids (see photo, right) and into the jar of beetles, being careful not to

Note the opening in the distribution jars' lids. To speed releases, simply remove the entire lid.

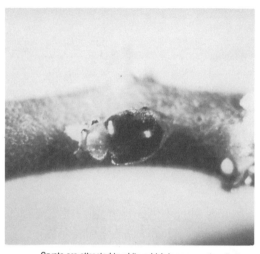
Crypts are attracted to white, which is one way they find their prey, like the beetle in this photo.

injure any of them; the beetles will climb up the pole to the other end [if it's not too hot, they may fly off the pole (but should still make it to the foliage — eventually)], which should be tucked into the tree's foliage. We've told some of our customers of this technique and encouraged them to try it. They did and reported excellent results. We now suggest everyone try it. Thanks Linnaea!

Last advisory: don't wear white clothing when you're releasing *C. montrouzieri*. They are attracted to white and light colors and may find you irresistible (see photo, below). Also watch your sticky traps, if you're using them. If you're catching too many beetles, remove the traps from the site or set them out for only 2-3 days per week.

Usages— Where can't they be used would a more appropriate question. We've seen the successful implementation of this species in just about every conceivable situation. All with equal success.

Rates—

PRVNT: N/A
CTRL-L: 2-4 / YD. - TRI-WKLY - 2-3X
CTRL-M: 4-6 / YD. - TRI-WKLY - 2-3X
CTRL-H: 6-8 / YD. - BI-WKLY - 2-4X
MAINT: 2-3 / YD. - MTHLY - INDEF
GRDN: 20-60% of rates above.
ACRE+: 3-20% of rates above.
COMMENTS: Large scale use is normally in the southern states. Usually in citrus.

Pricing information—
The *C. montrouzieri* beetles covered in this section can be obtained from The Green Spot by calling 603/942-8925. Detailed release instructions are provided with every order. Here is the current, industry-average pricing for 1998...

Item no. CCM1C ... 100 C. montrouzieri adults ... 1 unit = $22.40

Item no. CCM1C-4 ... 100 C. montrouzieri adults ... 4+ units = $20.72 ea.

Item no. CCM5C ... 500 C. montrouzieri adults ... 1 unit = $93.60

Item no. CCM5C-4 ... 500 C. montrouzieri adults ... 4-7 units = $81.90 ea.

Item no. CCM5C-8 ... 500 C. montrouzieri adults ... 8+ units = $75.76 ea.

Photo by M. Herbut, courtesy of Applied Bio-nomics, Ltd.

The Multicolored
Asian Lady Beetle a.k.a.
The Halloween Ladybug

Harmonia axyridis
(har-MONE-ee-ah ax-eh-RIDE-uss)

Description— Many of you are already pretty familiar with these beetles. Especially those of you along the eastern seaboard and in the pacific-northwest. *H. axyridis* are the beetles which have, for the past few years, been invading people's homes for the sake of comfy, protected overwintering (see Fig. 3, below).

Now don't be alarmed, we're offering only small quantities of these beetles for interior uses and, unlike their wild brothers and sisters, these beetles are not wild-collected, they're insectary reared just for the purposes specified.

Halloween ladybugs, with their shiny, rotund red, yellow or orange bodies may potentially clean up large populations of **scale insects**. Moreover, as most beetles are, *H. axyridis* can be very opportunistic and will eat pests other than scales: **whiteflies, mealybugs, insect eggs** and, in profound numbers, **aphids**. Moreover, these beetles may consume **European red mites**. We *do* recommend trying these beetles for other pests, especially aphids. We believe control of other pests *may* be obtainable with typical releases. Coincidental cross-predation should be regarded as a bonus of the application and *may* be more. Mass aphid consumption, in addition to scale insect control, should be expected (see photo, next page).

Harmonia axyridis are shipped in small lots as pre-fed, pre-mated adults. Some popular prey of these beetles is suspected to include: the **scale insect species** listed as hosts

FIG. 3

HAVING PROBLEMS WITH BEETLES OVERWINTERING IN YOUR HOME?

The small number of beetles we're offering will not significantly add to the problem some people are having with Halloween ladybugs overwintering in their homes. If you are having a problem with these beetles it is due to primarily wild populations. Regardless, the way to handle the invading beetles is as follows:

1) With a clean filter bag installed in your vacuum cleaner, suck up the clusters of beetles and a) put them in a greenhouse or interiorscape or, b) put them outside in a protected location such as a woodshed, etc. There's no need to kill them. They won't really cause any damage to life or property (even though great numbers of beetles may produce stains if disturbed).
2) Prevent their entry by caulking cracks, installing window screens, and keeping the doors of the house closed (go ahead, tell your kids "just one more time"), etc.

for *Aphytis melinus* and *Metaphycus helvolus* (both listed in this section) and other related species as well, plus a **wide assortment of aphids** (every species listed in the aphid controls section).

H. axyridis adult munching on aphids which
have taken over a nasturtium.

Life-style— These large and beautiful 8 mm. Asian, adult female beetles lay their eggs amongst pest colonies, averaging up to 700 eggs each; one female beetle laid over 1500 eggs in one test! The eggs hatch into black and orange larvae (see photo, facing page) similar to that of *Hippodamia convergens* (see aphid controls). These, too, are fierce predators, growing up to 1.5 cm. long and consuming a great number of pests.

The life-span of these predators is roughly 3-3½ weeks in their immature stages, then *up to* three years as adults (probably less under most conditions). The conditions for optimum performance will be between 70-85°F with a relative humidity of around 70%. But these are *optimum* conditions, and not necessarily a prerequisite of successful implementation. Please note, how-ever, cooler temperatures will hamper reproduction and development a certain degree.

Benefits— These predators should provide outstanding results in the long-term. Canadian interiorscape trials of scale insect control have proven quite successful thus far. Our summer, 1996 and 1997, trials were quite favorable. We were able to make numerous recoveries, especially larval.

Another significant benefit to using these beetles is the way they're supplied: insectary-reared. pre-fed, pre-mated. Oviposition or egg-laying can poten-tially occur very soon after release. These same beetles, if wild collected, would probably perform poorly unless released much the same way *H. convergens* are.

Drawbacks— We have little practical experience with these predators. However, past and present research is somewhat promising. We strongly suggest growers and interiorscapers try this species, but to do so not out of desperation.

In 1997, the first year these beetles were offered commercially, we discovered that they tend to be flightier than originally suspected. The grower should be thinking long-term when employing these beetles. It is very possible, in anything but a

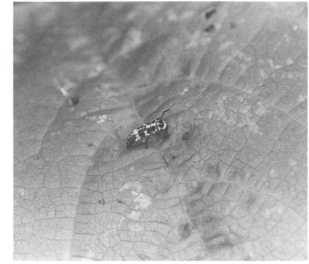

H. axyridis larva shown on bean leaf.

closed system, that few beetle recoveries will be made. Some folks who have trialed them said that they only retained one or two "token" beetles.

These beetles *may* be a bit on the pricy side. This will probably be the case until production peaks. This hinges on steady demand. We suspect that the results may far outweigh the price issue. Plus the application rates are nowhere near where they are for *H. convergens* (see Rates).

Scouting— Adult beetle presence, larval presence, reduction of pest numbers and new, clean plant growth. These are all signs that Halloween ladybugs are hard at work.

We have nothing more to offer at this point in time.

Advisories— Aside from misting the site with water before releasing and doing so in the evening (sometimes not necessary in interiors), there are other things you can do to ensure the maximum number of beetles perform. Flowering, pollen producing plants are a big plus (a seed mix like Bioblend, discussed in this manual, may be useful to these predators).

Pollen isn't the only thing these beetles will eat. They will also consume

Both photos, facing page and above, by M. Cherim

aphid and scale insect honeydew. A product such as Biodiet™ (a honeydew substitute) may help encourage the beetles.

Ants, if present, might need to be controlled. They will defend pests from predators and parasites to protect their honeydew/excrement food, gag! Use barrier products or boric acid products to control the ants.

You may want to try releasing these beetles into trees the way we are suggesting to try for Crypts: Find a long bamboo or wooden pole with an approximate diameter of three-sixteenths to one-quarter-inch; insert one end of the pole through the pre-manufactured hole in the bottle's lid and into the bottle of beetles, being careful not to injure any of them; the beetles will climb up the pole to the other end [if it's not too hot, in which case they may fly off the pole (but should still make it to the foliage — eventually)], which should be tucked into the tree's foliage.

Yellow sticky traps *may* present a problem with these beetles. We are unsure of this and ask that you exercise cautious observation when the traps are in use. If predator numbers on the traps get too high, remove the traps or use them for only two or three days per week.

Based on 1997 results, plan on making your H. axyridis releases after nightfall to minimize the fly-off and increase the likelihood of retention.

Usages— We're not certain of every possible use. We suggest trying them in greenhouses and interiorscapes mainly. However, orchard and garden uses may yield satisfactory results, especially if being used for aphids.

Rates—

```
PRVNT: N/A
CTRL-L: 1-2 / YD. - TRI-WKLY - 2-3X
CTRL-M: 2-3 / YD. - TRI-WKLY - 2-3X
CTRL-H: 3-4 / YD. - BI-WKLY - 3-4X
MAINT: 1-3 / YD. - QTRLY - INDEF
GRDN: 60-85% of rates above.
ACRE+: 10-35% of rates above.
COMMENTS: The release rates shown above
are suggested until further data is available.
```

Pricing information—
The H. axyridis beetles covered in this section can be obtained from The Green Spot by calling 603/942-8925. Detailed release instructions are provided with every order. Here is the current, industry-average pricing for 1998...

Item no. CHA50 ... 50 H. axyridis adults ... 1 unit = $26.25

Item no. CHA50-6 ... 50 H. axyridis adults ... 6+ units = $23.62 ea.

Item no. CHA1C ... 100 H. axyridis adults ... 1 unit = $44.10

Item no. CHA1C-4 ... 100 H. axyridis adults ... 4-7 units = $40.79 ea.

Item no. CHA1C-8 ... 100 H. axyridis adults ... 8+ units = $36.71 ea.

The Soft-Scale Parasitoid

Metaphycus helvolus

(MET-ah-fii-kuss HEL-vole-uss)

Description— These gold, 1.3 mm. mini-wasps are best used as preventive agents. They can also be used to tackle and maintain minor to heavy infestations of several soft scale species. And, if established, they can adequately protect a planting for a season or, perhaps, more in certain situations.

M. helvolus, which are shipped as pre-hatched adults, are the product of choice when soft scales that can host this parasitoid are present. (These wasps are shipped with Biodiet™ on the lid of their jar, but it is provided merely as a moisture source. These parasitoids do eat the stuff, but it is not recommended for supplemental usage at your release site.)

Some of the species which can be controlled with these parasitoids include the **black scale (their favorite), Mexican black and hemispherical scale insects (*Saissetia oleae, S. miranda* and *S. coffeae*, respectively); nigra scale (*Parasaissetia nigra*); the *brown soft and citricola scale insect species (*Coccus hesperidium* and *C. pseudomagnoliarum*, respectively); and other economically important species.** (*Brown soft scale control has been debated. Some researchers say it can't be done. Yet, in several forced situations, we've seen them work quite well — even at the University of New Hampshire, where the results were confirmed. We suggest growers perform their own trials.)

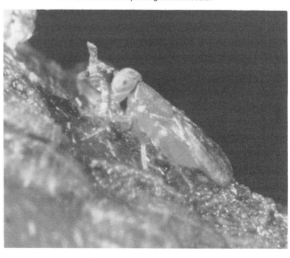

M. helvolus adult ovipositing in a soft scale.

Life-style— These parasitoids, work by laying eggs inside scale insects (see photo, right). They will lay up to 100 eggs, killing 5 to 25 pests as they do. They will host feed on many more. The wasps' larvae which

Photo by M. Badgley, courtesy of Buena Biosystems, Inc.

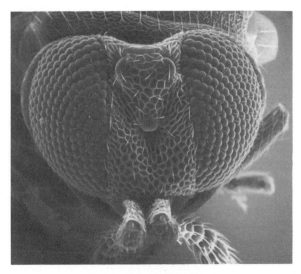

Like Aphytis melinus, M. helvolus search for suitable hosts with their antennae.

hatch from the eggs, slowly weaken and kill the scales from within (*endo*parasitism). New parasitoids hatch out of the empty scale cadavers after they pupate.

The life-span of these parasitoids is roughly 42 days in their immature stages, then around 2 months as adults. The conditions for optimum performance will be between 73-87°F with relative humidity of around 50%. But these are *optimum* conditions, and not necessarily a prerequisite of successful implementation. Please note, however, cooler temperatures will hamper reproduction and development a certain degree.

Benefits— These parasitoids, with the right host, in the right conditions, are superb performers.

M. helvolus are superb curative agents, thus offering growers a potential money-saving tool (they *can* be more economical than pesticides). Additionally, they can establish themselves in nearly any situation. Once established, growers might be able to reduce the size of, or curtail, future releases due to the presence of on-site wasps: another money-saver.

Lastly, these parasitoids, being shipped as adults, offer the benefit of fast oviposition or egg-laying, plus a lot of host-feeding!

Drawbacks— They won't overwinter in cold climates; they can be difficult to scout (see next); they are attracted to lights, light colors and sticky traps; and ants and honeydew must be under control. Whew! These drawbacks offer some interesting challenges at times.

Fortunately that's it. Not! One more thing: they're a bit pricy, too. The latter, though, can be debated by many of our customers since these mini-wasps have such potential. We agree.

S.E.M. photo by N. Cherim

Scouting— Reduction of pest numbers (empty scales and lack of crawlers) is a sure sign of success.

The scout can also usually find scale insect cadavers with an exit hole in them. This is a very positive sign.

Trying to spot the wasps themselves, say, two weeks after a release, would be an exercise in futility.

A reduction in natural honeydew quantities and clean new growth are other notable indicators of parasitoid success.

Advisories— *Metaphycus helvolus* eat honeydew, but don't particularly like to get the stuff on them. If honeydew is excessive, these parasitoids will spend most of the time cleaning themselves.

Ants are another serious issue related to honeydew. They must be controlled. Use barrier products or boric acid products to control the ants, if they are a problem.

Yellow sticky traps should be removed prior to releasing these mini-wasps. To monitor for thrips, use blue traps. If yellow traps must be used for whiteflies, etc., hang them for only two to three days per week.

Usages— Greenhouses and interiorscapes are the settings where these parasitoids are most often employed by our customers. However, in the right environment, these wasps can be used with great success outdoors. After all, they *are* reared especially for the California citrus industry.

Rates—

```
PRVNT: 1-2 / YD. - QTRLY - NEED
CTRL-L: 1-2 / YD. - MTHLY - 1-2X
CTRL-M: 2-3 / YD. - TRI-WKLY - 2-3X
CTRL-H: 3-4 / YD. - TRI-WKLY - 2-4X
MAINT: 1-2 / YD. - MTHLY - INDEF
GRDN: 60-85% of rates above.
ACRE+: 15-35% of rates above.
COMMENTS: Spring releases are recom-
mended outdoors. Indoors, use anytime.
```

Pricing information—
The *M. helvolus* wasps covered in this section can be obtained from The Green Spot by calling 603/942-8925. Detailed release instructions are provided with every order. Here is the current, industry-average pricing for 1998...

Item no. CMH5C ... 500 M. helvolus adults ... 1 unit = $61.42

Item no. CMH5C-4 ... 500 M. helvolus adults ... 4+ units = $56.81 ea.

Item no. CMH25C ... 2500 M. helvolus adults ... 1 unit = $262.75

Item no. CMH25C-2 ... 2500 M. helvolus adults ... 2+ units = $243.00 ea.

The Singular Black Lady Beetle a.k.a. The Scale Destroyer

Rhyzobius = *Lindorus lophanthae*
(Ry-ZOBE-ee-uss =Lin-DORE-uss loh-FAN-thay-ay)

Description— Like *Harmonia axyridis*, our new scale/aphid predators shown in this section, *R. lophanthae* are ladybug beetles.

These lady beetles, with their small, fuzzy black bodies and dull-orange head/thorax region, love scale insects. They can clean up large populations. And, as most beetles are, *R. lophanthae* are very opportunistic and will eat **pests other than scales: mealybugs (their crawlers or immature forms), insect eggs, etc.** We do NOT recommend using these beetles for other pests, though. We believe control of other pests might not be obtainable with typical predator releases. Coincidental cross-predation should be regarded as a bonus of the application and nothing more.

With scale insects, though, these beetles have proven themselves worthy. Our customers have reported good results with their use. Especially our interiorscaper and botanical garden customers.

R. lophanthae are shipped as pre-fed, pre-mated, insectary-reared (for the citrus industry) adults. (These beetles are shipped with Biodiet™ on the lid of their jar, but it is provided merely as a moisture source. These predators eat very little of the stuff, so it is not recommended for supplemental usage at your release site.) Some popular prey of these beetles include: the **scale insect species** listed as hosts for *Aphytis melinus* and *Metaphycus helvolus* (both listed in this section) and other related species as well.

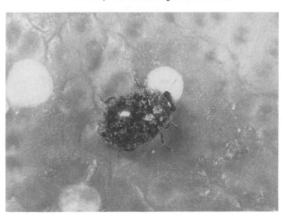
Adult R. lophanthae attacking an armored scale.

Photo by M. Badgley, courtesy of Buena Biosystems, Inc.

They do seem to do

well with controlling many armored scale insects and black soft scales (see *M. helvolus*). We're told that they are also good **purple and California red scale predators (*Lepidosaphes beckii* and *Aonidiella aurantii*,** respectively).

Life-style— The smallish 2.5 mm. (see photo, facing page) adult female beetles lay their eggs *underneath* scale insects — hundreds of them. The eggs hatch into small gray larvae. These, too, are fierce predators, growing up to 3 mm. long and consuming vast numbers of small scale crawlers and eggs (they'll always eat the youngest, most tender morsels first).

The life-span of these predators is roughly 3 weeks in their immature stages, then 5-8 weeks as adults. The conditions for optimum performance will be between 73-87°F with a relative humidity of around 65%. But these are *optimum* conditions, and not necessarily a prerequisite of successful implementation. Please note, however, *considerably* cooler temperatures will hamper reproduction and development a certain degree (see Benefits).

Benefits— These beetles have been observed going about their business in temperatures as low as 40°F. We doubt they were laying eggs, though.

R. lophanthae have an excellent establishment ability in the right environment.

Singular black lady beetles are pretty tough — though you might not think so after seeing how small they are. They have an excellent ability to chew right through the hard covers of both armored and soft scales.

Drawbacks— They are very difficult to scout. We only have a handful of customers who've had success with this task.

They can't handle too much honeydew (scale poop). If they are going to be used on soft scale insects, check out the honeydew situation first.

Scouting— Look for scales which look like they've been partially eaten. Finding the adults or larvae can be fairly exhaustive work.

Clean new plant growth and natural honeydew reduction may also be apparent scouting indicators.

Advisories— Aside from misting the site with water before releasing and doing so in the evening (sometimes not necessary in interiors), there are other things you can do to ensure the maximum number of beetles concentrate on the pests at hand. Flowering, pollen producing plants are a big plus.

Unscrew the cap or use the small hole on top to release the beetles in this 100 count unit. Yes, we know this photo is old: "L." lophanthae, oh, well.

Ants, if present, should be controlled. They will defend scales from predators and parasites to protect their honey-dew/excrement food, ick! Use barrier products or boric acid products to control the ants if it is necessary.

You may want to try releasing these beetles from their jar (see photo, inset) with a long pole if you're going to be doing so in trees, as described under *H. axyridis*' Advisories (also under Scale Insect Controls).

These beetles may be attracted to light colors, so watch your sticky traps, if you're using them. If you're catching too many beetles, remove the traps from the site or set them out for only 2-3 days per week.

Usages— Greenhouses, interiorscapes, southern orchards and citrus. Anywhere conditions are right and food is plentiful!

Rates—

PRVNT: N/A
CTRL-L: 2-4 / YD. - TRI-WKLY - 2-3X
CTRL-M: 4-5 / YD. - TRI-WKLY - 2-3X
CTRL-H: 5-7 / YD. - BI-WKLY - 2-3X
MAINT: 1-2 / YD. - BI-MTHLY - INDEF
GRDN: 35-55% of rates above.
ACRE+: 5-20% of rates above.
COMMENTS: Large outdoor releases should be confined to the southern states.

Pricing information—
The *R. lophanthae* beetles covered in this section can be obtained from The Green Spot by calling 603/942-8925. Detailed release instructions are provided with every order. Here is the current, industry-average pricing for 1998...

Item no. CRL50 ... 50 R. lophanthae adults ... 1 unit = $39.00

Item no. CRL1C ... 100 R. lophanthae adults ... 1 unit = $67.60

Item no. CRL1C-6 ... 100 R. lophanthae adults ... 6+ units = $60.84 ea.

Item no. CRL5C ... 500 R. lophanthae adults ... 1 unit = $281.38

Item no. CRL5C-4 ... 500 R. lophanthae adults ... 4+ units = $267.31 ea.

Photo by M. Cherim

soil pests

Soil pests are comprised of species from several orders and families.

Soil pests — now there's a broad category. It consist not only of critters which ordinarily live in the soil full-time, like phytophagous or plant-eating nematodes, but others as well. For example, fungus gnats and certain beetle species spend just some of their lives in the soil. To further clarify, let's take a closer look at a couple of these part-timers.

The fungus gnat adult, above, is being viewed from overhead. Hopefully this photo will clearly show the 'Y' shaped wing veins which help us identify this pest.

Below, are the fungus gnat's larvae in the moist, comfortable surroundings: your media.

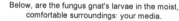

Fungus gnats are small gnat-like insects (see photo, above) which, in their adult stage, are a nuisance pest. These gnats lay their eggs on a moist growing media (soil, peat-based products, rockwool, etc.). These eggs hatch into small clear/white, black-headed larvae (see photo, right) — sometimes sporting a black alimentary canal stripe full of yuk

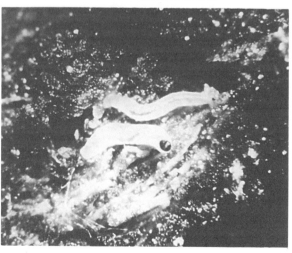

Both photos by M. Herbut, courtesy of Applied Bio-nomics, Ltd.

— which attack minuscule fungal particulates, plant roots/root hairs and the such. Doing so, they also contribute to, and, in some cases, cause, soil diseases. This happens because as they move about through the media, they may contact various pathogens, expose the cell structure of the roots, and generally weaken and stress the plant.

The half-inch metallic, bronze-green Japanese beetles (see photo, below) are another fine example. Around July they'll come out of the ground as adults. They'll feed — causing mass destruction — mate and lay eggs. The eggs hatch in the soil, and the resulting larvae begin to feed on turf- and plant-root systems, fattening themselves for the winter. In the spring the continue to feed on roots and fatten themselves up. They pupate, while still in the soil, then emerge as adults. And this is where we came in.

Adult Japanese beetle pigging out on a flowering crab leaf.

Fungus gnats and Japanese beetles are just a couple of examples. Cucumber beetle (see photo, facing page) and, flea, Oriental and June beetles are others. In fact, 90% of all insects spend at least a small portion of their lives dirty. What a great killing field.

Now we have to ask ourselves if we're ready to control these pests — and how? Soil sterilization and/or fumigation may come to mind, which is fine, however, consideration for the other, non-harmful, soil-inhabiting organisms should probably be taken into account. Numerous microorganisms, necessary for life on earth as we know it, inhabit the same soil as the target pests previously discussed. Should we chance killing those with either method? How about the wildlife above ground which may be affected by below ground fumigation? You're in a greenhouse growing plants on soilless media' does this apply to you. Does runoff mean anything?

So many pests causing yellowing patches of lawn in their larval stages, only to become adults and destroy our weakened plants above ground; such variety.

Both photos, above and facing page, by M. Cherim

Chemicals certainly will give you a broad-spectrum kill, but, considering the consequences, an alternative would probably be preferred. (See soil pest controls for the solution to the problem, next.)

The controls in the next section — non-plant-eating, but bug bugging, nematodes — are all-powerful and will impact a large number of the pests you're concerned with. However, you'll still want to know what you're dealing with, how it overwinters and where. Start in the summer if you have no historical records to tell you what's what. Observe and identify the many critters found in your growing area. See the plants they like most? Those are the plants they will likely overwinter under (or simply go through metamorphosis only to reemerge in the same season). If there's turf underneath, peel it back to a depth of 2-3 inches. The plump whitish, C-shaped things are probably the grubs or larvae of the beetles with which you're concerned.

Perhaps you're in an interiorscape, and you are pretty sure the little buggers flying around are fungus gnats. Watch them to see which pots or planter boxes they're frequenting. Scrape a 1-2 inch layer of media off the top. The tiny larvae should be visible if present.

The striped cucumber beetle is one tough pest to contend with. They only spend their pupal stage underground.

Dealing with pests in the soil requires extra special attention on the part of the pest control manager. Remember, in the case of these pests, the treatments are applied directly to mother earth.

For more detailed information and recommendations concerning soil pests and ways to effectively control them, you are invited to call The Green Spot, Ltd. at 603/942-8925. ▓

soil pest controls

The Parasitic

Nematodes

Heterorhabditis bacteriophora
(het-ter-rore-hab-DITE-iss bak-teer-ee-OFF-for-ah)

Steinernema carpocapsae
(STINE-er-neem-ah kar-poh-KAP-say-ay)

Description— These nearly-microscopic nematodes live in the soil and spend their lives exploiting the larvae and/or pupae of over 200 organisms for their own gain. They are entomogenous (en-toh-MAH-jen-us), meaning they develop on or within an insect.

H. bacteriophora (Hb), which are offered as live 3rd stage juveniles, (J-3) *infective* stage, on sponges (see photo, right), are the product of choice when your problems concern lethargic, deeply planted pests: **Japanese beetle [larvae] grubs (*Popillia japonica*)**, for example. The reason: these nematodes boast a deep-moving (1-7"), active-hunting characteristic which make them superior to many other species for the purposes specified.

S. carpocapsae (Sc), which are also offered as live 3rd stage juveniles, seem to be the product of choice when your problems concern more active, shallowly occurring pests: **fungus gnat or mushroom fly larvae (i.e., *Lycoriella mali* and *others from different genera)**, etc. The reason: these nematodes feature a shallowly-waiting (0-2"), ambush-hunting characteristic

---

FIG. 4

Fungus Gnat Gnews

Fungus gnats, or mushroom flies as they're sometimes called, are members of one of three genera (a taxonomical grouping of insects by type, traits, etc.): Lycoriella, see above, Sciara (Sciarid flies) and Bradysia.

The black/gray adults look like small 2-3 mm. mosquitoes. They can be a nuisance, especially in interiorscapes. Trapping them is easy with yellow sticky traps.

The larvae look like small, elongate 2 mm. grubs with translucent milky-white bodies sporting shiny black head-caps. They are a major concern of greenhouse and mushroom growers, and interiorscapers everywhere. They feed on fungus as their name implies, and ⟩⟩⟩

which make them superior to many other species for the purposes specified. (*For more on fungus gnats, see Fig. 4, bottom of page.)

Sponge packaged nematodes like the J-3 Max products offered by The Green Spot, are very common. Not all are reared in-vivo though.

A mixture of *both* species is useful in a single application against a variety or mix of pests. An Hb/Sc mix is often used for fungus gnats in larger containers. The Hb nematodes will work the deeper ground, say around the drainage holes, while the Sc nematodes guard the surface soils.

Another useful application for a nematode *mix* is to combat several **cutworm species, black and variegated cutworms (*Agrotis ipsilon* and *Peridroma sauci*, respectively), for example.

Other pests which can be controlled, or at least impacted, with one or both of the species include: **apple maggot pupae (*Rhagoletis pomonella*); black vine weevils (*Otiorhynchus sulcatus*); cabbage root maggots (*Delia* spp.); carrot rust flies (*Psila rosae*); carrot weevils (*Listronotus oregonensis*); spotted and striped cucumber beetles (*Diabrotica undecimpuncata howardi* and *Acalymma vittatum*, respectively); flea beetles (*Podagrica uniforma*); many fruit borer species (*Synanthedon* spp.); June/May**

Photo by M. Cherim

also on root hairs, and if gnat-initiated root rots occur, as they probably will, the larvae will sometimes invade the larger roots, crown and main stem. Their feeding will cause the leaf disfiguration of many plants if left unchecked. Trapping the larvae is easy with quarter-inch slices of raw potato placed on the soil's surface, checked weekly, at least, for burrowing larvae.

The Bad Gnews: as long as growers use peat products, composts, etc., the pests will be a part of life.

The Good Gnews: Hb and Sc parasitic nematodes and Hypoaspis miles (soil dwelling mites listed in detail under Thrips Controls) can provide superb control. Nematodes for the quick fix or for low container-volume situations, H. miles for the long-haul on floors, in planters and larger containers. (Shiitake mushroom growers, see H. miles.) ✄

beetles (*Phyllophaga* spp.); leafminer spp. pupae (i.e. *Liriomyza trifolii*); northern corn root worms (*Diabrotica longicornis*); onion maggots (*Delia antiqua*); potato tuber worms (*Phthorimaea operculella*); squash vine borers (*Melittia cucurbitae*); strawberry root weevils (*Otiorhynchus ovatus*) wireworms (*Limonius* spp.); and many more economically important critters.

The are also other nematode species available to bio-control practitioners: *Steinernema feltiae*, *S. riobravis* and others. All boast a special, slight advantage over the others but, in actuality, for most practical purposes, the inexpensive Hb/Sc nematodes get the job done. However, there are some new strains being introduced which allegedly attack hosts which are unaffected by the common Hb/Sc nematodes; host such as other nematodes [?] and slugs. One example is *Phasmarhabditis hermaphrodita*; it attacks slugs.

Life-style— Both nematode species physically enter the host [grub] and kill from within (*endo*parasitism). Entry is gained via the host's mouth, anus or spiracles (breathing holes located along the sides of an insect). The Hb nematode, with its well structured and muscular, beak-containing head also has the ability to enter through the host's soft, side-wall tissue. The Sc nematode has a more slender head (see photos, below).

Before they can enter the host, though, it must be found. Both nematode species accomplish this by "tasting" or "sampling" the air [soil air] for the "flavors" of carbon dioxide, mainly, and also methane gas residues and warmth. The Sc nematodes will actually "stand" on the soil's surface and waver back and forth — tasting, testing, sampling. If they lock onto a scent

The blunt, muscular head of Hb nematode (left), as compared to the sharply tapered head of the Sc nematode (right). Both are magnified 2,000 times for these scanning electron micrographs.

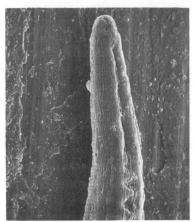
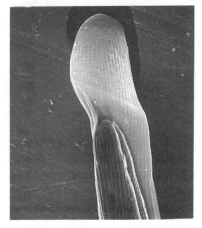

Both S.E.M. photos by N. Cherim

they will travel a short way to it, lie in wait for it or they might *leap* to it. In our lab we've seen Sc nematodes *actually leaping up to a centimeter* in a petri dish containing moist peat and vermiculite. *And That Was Cool!*

In the case of the Hb nematodes, they simply search the soil, tracking down their host, slithering along the film of moisture surrounding the medium's particulate matter, sampling as they go.

Once inside the host, the nematodes settle in, shed a protective cuticle, mature, and begin feeding, defecating and reproducing. In their fecal matter lives a symbiotic bacterium enjoying the perfect harmony of interdependency with the nematodes. The bacterium present in the fecal matter of Hb nematodes is *Photohabdus luminescens*, in the Sc's is the bacterium strain *Xenorhabdus nematophilus*. Both types poison the host's [grub] blood, thus killing it. The nematodes don't actually do the killing, they just make it all possible.

The life-span of these nematodes is roughly 8 weeks for Hb and 15 for Sc, all together. They progress through 4 immature stages J-1 thru 4. 1-3 they live in-vivo (inside their host), J-3 they're ready to hit the road, J-4 and adulthood are spent in-vivo in their *new* host. Their life-cycle cannot progress beyond stage J-3 until they find a new host. This can artificially lengthen their live, but for no good cause.

The conditions for optimum performance will be soil temperatures between 65-70°F for Hb and 60-70°F for Sc. Both prefer the soil, soilless medium, rockwool, etc., to be slightly moist. But these are *optimum* conditions, and not necessarily a prerequisite of successful implementation. Please note, however, cooler and warmer temperatures, and insufficient moisture levels, will hamper reproduction and development a certain degree.

Temperature can play an interesting role with nematodes, one that can be manipulated a certain extent. For example: Hb's bacterium requires a constant temperature of 60°F+ to break dormancy. However, Hb nematodes are active at 45°F, barely. The same's true for Sc nematodes: the bacterium needs 55°F+, and the nematode is active at 40°F. Additionally, soil temperatures above 90°F are too high. The nematodes can be put into the ground early, as long as no more freezes are expected; anytime indoors. (Their bacterium will become active when the conditions are right.) In fact, our nematodes should be used whenever the target pest is in the ground in a susceptible form, usually larval, sometime pupal. Some releases should be made early in the season to catch spring pupating grubs, or later in the year when adults are laying soon-to-hatch eggs.

Benefits— Nematodes are very inexpensive to use and provide fairly fast results. The first host death will occur within 48-72 hours of host penetration for Hb nematodes, and 16-24 hours for Sc. Results can usually be maintained for 6-15 weeks. [Not all will work right away, though (see Drawbacks)] Using new planting medium will probably spark the need for additional applications, though, especially if a peat-based product is being used. Other causes of shorter performance periods can include: allowing containerized plants to dry, bone-dry, completely; ultra violet light (UV) exposure (see Advisories); and the use of a nematicide. (Nematodes are compatible with most other substances.)

Nematodes are very versatile. They can be applied a number of different ways, through both automated equipment and manually. They can be put through just about any irrigation setup; they'll withstand up to 300 p.s.i.

Versatility extends beyond the flexibility of their application. The "little worms with a big attitude," as we sometimes refer to them, attack a huge segment of the pest species which bother growers. This one feature allows them to often be used successfully in cases where everything else has failed. Sometimes a grower will encounter a situation where we haven't a clue what to offer as a remedy — and chemicals have proven themselves worthless. We'll do some research with the grower and find out the pest spends part of its life underground (as 90% do). We then might suggest the grower explore the nematode option. He or she does, and sometimes we all win! One arborist wanted to try them against a species of borer by spraying them on the trunks of large trees. Apparently the nematodes liked the borers, their holes and the rough bark of the trees, because they performed much better than expected.

Please note, however, as much as their versatility is beneficial, it can also be a detriment (see Drawbacks).

Most nematodes can be stored. In-vivo nematodes typically for 2-3 months in the refrigerator. Some other formulations may be in long-term suspended animation for up to 6 months or more sometimes. Short-term storage does not hurt *athletic,* healthy-to-begin-with nematodes. They simply live off of their lipids, or fat cells. Long-term storage may have undesirable effects to the overall quality of nematodes, but it depends a lot upon their initial pre-storage health (see Advisories).

With healthy nematodes, buying in quantity and storing can be to your advantage. You might be able to save on shipping, handling and product costs, if volume breaks are offered and met. If repeatedly applied, nematodes will provide maximum results.

Lastly, nematodes will not harm people or animals, not even earthworms.

Drawbacks— These nematodes don't seem to establish themselves to such a degree as to continue getting pest control two years down the line in sod. The Hb species *will* bear up to 200,000 young 'uns beginning in as little as 9 days, the Sc species can have the same number beginning in a mere 7 days. Despite this, though, reapplications are usually necessary annually, or even at shorter intervals for containers (see Rates), depending upon pest pressure, conditions, and attractant/host material. Another factor affecting this is the fact that these nematodes don't all go to work at the same moment. With 100% live and heathy nematodes, the most to immediately go to work, typically, is 40%. Mother Nature always thins things out. She tries all the time to even the playing field. She's looking to achieve eventual equilibrium — balance. This is why the nematodes can't remain in the numbers we need for sustained control, not without physically barring the pests from the plant material or removing it all together. Good thing nematodes are reasonably priced!

Hb/Sc nematodes have a wide host range. To say the least! This is true because the nematodes are nonselective. They shouldn't be used with the aphid midge, *Aphidoletes aphidimyza*, (see aphid controls) for this reason (they *are* okay with the soil-dwelling mite, *Hypoaspis miles*). Other beneficials, some natural ones, and some of the other commercially available species, can sometimes fall victim to nematodes, but not to an extent which may render a program ineffective. This should only be a minor concern (if at all), but it's one you should be aware of.

Some pests we can positively say they don't parasitize, which in itself is a drawback, are other nematodes, the pernicious plant pests: lance, root-knot and dagger nematodes, for example.

The application of nematodes can be an arduous task without some sort of equipment. Doing it by watering can is a very effective method, but it's time-consuming. Fertilizer injectors, boom sprayers, etc., make it much easier. For small outdoor sod applications, certain hose-end sprayer work like a charm and are of little cost.

Scouting— Parasitized grubs will be, 1) dead, 2) partially deteriorated (depending on how long after initial host entry they were discovered), and/or 3) discolored. The victims of the Hb nematode will be tinged orange-red, and those of the Sc will be only be a slightly discernible yellow.

But before anything can be parasitized, the nematodes must be alive and ready for action. We have an easy way that you can test them when you receive your order:

1) Remove the sponge from the bag.

2) Apply a tiny amount (a pencil-point amount) of the nematodes (the off-white to yellow material on the sponge) to the plastic bag in which they came.

3) Add 1-2 drops of room temperature water to the nematode sample.

4) Wait 2 minutes.

5) Place the plastic bag against a smooth, dark surface.

6) With a hand-lens, and sometimes with the naked eye if conditions and your eyesight are perfect, you should be able to detect movement in the droplet. With 20x and up, individuals should be seen "swimming." (See photo, below).

7) This process should work with any nematode product. The difference will be how the nematodes are collected and suspended, i.e., a granular material will have to be soaked in water to allow the nematodes to swim — or float — freely. Moreover, concerning non-in-vivo nematodes, a resurrection period of up to 72 hours may be needed before movement [life] is noticed.

Advisories— As previously explained, repeated drying and re-hydrating of nematodes can adversely affect them. The same holds true for UV exposure. They're soil-dwelling critters and can't handle a good tan. Direct sunlight will first sterilize the nematodes (this takes only about 7 minutes), then it will kill them. It is for this reason that nematodes should be applied in the evening or night, or during overcast weather. Outdoors, a good time to apply these organisms is during a rainstorm. For one thing you won't have to worry about UV light and, second, you won't have to use so much water to properly wash them in the soil.

Here is an old micrograph of a gaggle of Sc nematodes. There are a lot of impurities with this batch of nemtodes — which will probably hamper storage. The manufacturer of the nematodes in this photo is unknown.

Micrograph photo by M. Badgley, courtesy of Buena Biosystems, Inc.

Speaking of water, if you wish to determine how much water you'll need, see Fig. 5, facing page. (But you don't *have* to saturate.)

An important advisory concerning nematodes is they way they are reared and supplied. Most nematode-based products today are made of nematodes artificially reared (in-vitro) on an artificial host and a diet of nematode chow. The nematodes are then placed in a state of suspended animation, where the vital functions of the organisms are all that remain. Clinically speaking, many of these *undead* nematodes are actually,

> **SOIL SATURATION TABLE**   Fig. 5
>
> Depending upon soil porosity, you'll need 'X' gallons of water to soak a square foot of soil to a depth of 'X' inches. 'X's =
>
> A P P R O X I M A T E L Y
> For a 2" saturation use 0.66 gallons,
> For a 4" saturation use 1.33 gallons,
> For a 6" saturation use 2 gallons,
> For an 8" saturation use 2.66 gallons, and so on.

well, dead. These typically come packaged by the billions in a gel or water dispersible granules. The vast nematode numbers compensate for the dead nematodes, and the sluggish behavior of the survivors. These products are designed for shelf-life so they can be sold through normal distribution channels and handled as an inventory item, complete with up-front manufacturers' profits.

Most of these are satisfactory products which work as indicated on their labels, just not as well or as economically as nematodes which are reared in-vivo, typically using an actual insect host: the larva of the wax moth (*Galleria mellonella*), a.k.a. the wax worm. One quality manufacturer allows their nematodes to hatch out of the host themselves and, by their own means, slither to a collection point — which confirms their vigor. This keeps the nematodes in peak condition, searching, tasting, sampling, hunting and killing. (This can't be done with an artificial diet and host). Some manufacturers induce suspended animation. They are sometimes not shipped alive — not really alive! In-vivo nematodes go to work immediately. They don't have to wait 24-72 hours to resurrect themselves. They can even be tested easily before you put them out to ensure the product is good (see Scouting).

Nematicides, as their name implies, kill nematodes. Products with this quality should not be used.

Usages— Greenhouses, fields, interiorscapes, orchards and gardens. We've seen the successful implementation of these species in just about every conceivable situation. This includes just about every conceivable media: soil/sod, peat-lite mixes, bark products, rockwool, etc.

Rates—

> **PRVNT:** 1-2000 / YD. - MTHLY - NEED
> **CTRL-L:** 2-3000 / YD. - TRI-WKLY - 2-3X
> **CTRL-M:** 3-4000 / YD. - BI-WKLY - 2-3X
> **CTRL-H:** 4-6000 / YD. - BI-WKLY - 2-4X
> **MAINT:** 2-3000 / YD. - QTRLY - INDEF
> **GRDN:** 100% of rates above.
> **ACRE+:** 85-100% of rates above.
> **COMMENTS:** Timing will vary from pest to pest.
> Nematode combos may need higher rates

Pricing information—

The *Sc/Hb, in-vivo* nematodes covered in this section can be obtained from The Green Spot by calling 603/942-8925. Detailed release instructions are provided with every order. Here is the current, industry-average pricing for 1998...

IN-VIVO REARED HB NEMATODES

Item no. CNHB1 ... 1 mil. Hb nematodes ... 1 unit = $10.80

Item no. CNHB1-8 ... 1 mil. Hb nematodes ... 8-15 units = $9.18 ea.

Item no. CNHB1-16 ... 1 mil. Hb nematodes ... 16+ units = $7.80 ea.

Item no. CNHB6 ... 6 mil. Hb nematodes ... 1 unit = $35.91

Item no. CNHB6-6 ... 6 mil. Hb nematodes ... 6+ units = $30.52 ea.

Item no. CNHB25 ... 25 mil. Hb nematodes ... 1 unit = $109.20

Item no. CNHB25-4 ... 25 mil. Hb nematodes ... 4+ units = $95.55 ea.

IN-VIVO REARED SC NEMATODES

Item no. CNSC1 ... 1 mil. Sc nematodes ... 1 unit = $10.80

Item no. CNSC1-8 ... 1 mil. Sc nematodes ... 8-15 units = $9.18 ea.

Item no. CNSC1-16 ... 1 mil. Sc nematodes ... 16+ units = $7.80 ea.

Item no. CNSC6 ... 6 mil. Sc nematodes ... 1 unit = $35.91

Item no. CNSC6-6 ... 6 mil. Sc nematodes ... 6+ units = $30.52 ea.

Item no. CNSC25 ... 25 mil. Sc nematodes ... 1 unit = $109.20

Item no. CNSC25-4 ... 25 mil. Sc nematodes ... 4+ units = $95.55 ea.

IN-VIVO REARED HB/SC NEMATODE COMBINATION

Item no. CNHSM1 ... 1 mil. Hb/Sc combo ... 1 unit = $10.80

Item no. CNHSM1-8 ... 1 mil. Hb/Sc combo ... 8-15 units = $9.18 ea.

Item no. CNHSM1-16 ... 1 mil. Hb/Sc combo ... 16+ units = $7.80 ea.

Item no. CNHSM6 ... 6 mil. Hb/Sc combo ... 1 unit = $35.91

Item no. CNHSM6-6 ... 6 mil. Hb/Sc combo ... 6+ units = $30.52 ea.

Item no. CNHSM25 ... 25 mil. Hb/Sc combo ... 1 unit = $109.20

Item no. CNHSM25-4 ... 25 mil. Hb/Sc combo ... 4+ units = $95.55 ea.

Item no. CNHSM250 ... 250 mil. Hb/Sc combo ... 1 unit = $468.00

Item no. CNHSM250-2 ... 250 mil. Hb/Sc combo ... 2-3 units = $397.80 ea.

Item no. CNHSM250-4 ... 250 mil. Hb/Sc combo ... 4-5 units = $367.96 ea.

Item no. CNHSM250-6 ... 250 mil. Hb/Sc combo ... 6+ units = $340.37 ea.

spider mites

Order: Acarina   Family: Tetranychidae & others

Spider mites are a very common problem, especially in dry areas with high heat. This may include greenhouses and homes during the dryer winter months, interiorscapes, etc. And although these eight-legged pests are small, they can be a serious and enduring problem. The most common spider mite attacking plants today is the two-spotted mite, Tetranychus urticae (see photo, below).

Normally, by way of animals, people or wind, the mite will hitch a ride and release itself in a new, promising location. This would be a stressed out plant of almost any kind, or a plant of particular interest to the mite — which could include any one of the vast number plants known to support two-spotted mites.

A newly relocated female mite will probably lay some eggs, if possible. These clear to milky 0.1mm. eggs, typically deposited to leaf undersides, will hatch into barely detectable clear larvae, which, in mere days, transform through various nymphal stages, changing color and size as they grow, to adulthood. As adults, they then have the capability of expanding the population — which they eagerly do.

The spots of the two-spotted spider mites, for which they were named, will sometimes appear as one, connecting like saddlebags over the back.

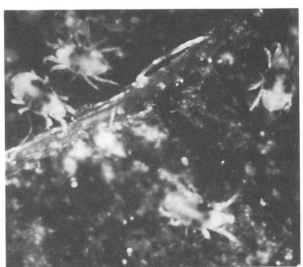

Photo by B. Costello, courtesy of Applied Bio-nomics, Ltd.

It is virtually impossible to detect mites on a plant before the plant shows some sign of the pests' presence. This puts the scout at an unfortunate disadvantage. However, with proper scouting at appropriate intervals, mites can be caught early, before reversal gets difficult.

Often, the scout

The cucumber plant's leaves in this photo show serious yellowing and speckling from spider mite feeding. The webbing is not visible, but most definitely present.

notices webbing as the first indicator. Unfortunately webbing is a sign of an advanced situation. The webbing is usually supported by the leaf's underside and midrib, which, in turn, creates a tent of sorts. This tent has a lot of activity going on inside: reproduction, egg-laying, development of immatures, drinking and dancing (okay, not the latter two, just wanted to see if you are still with us).

If at all possible, physically remove this webbing, it is an important part of controlling spider mites, regardless of how you'll finish the job. And since mites like hot, dry conditions, lowering the temperature and increasing the humidity (even through regular misting) will slow the reproductive pace of the mites. Washing with fairly high pressure water will also prepare the control site. A well-aimed water stream can seriously set the mites back simply by knocking them from the plant and destroying their webbing (and water streams are a good way to increase the humidity

Another indicator of a spider mite activity, before webbing is noticed, but when the infestation is still considered fairly high, but, perhaps, just in concentrations, is the speckling and yellowing of the leafs. As the mites suck the fluids from the leaf's cells, the cells dry and yellow from their emptiness (see photo, above). In the right conditions, typically in a structure, mites can be a serious pest. Tomato and cucumber growers, interiorscapers, and many others, suffering from a nagging perpetual spider mite problem.

On most crops, spider mites can be a pretty straightforward problem with a solution in sight. But only if the grower is willing to put-forth some effort in employing several tactics, and really sticking with them.

For more detailed information and recommendations concerning spider mites and ways to effectively control them, you are invited to call The Green Spot, Ltd. at 603/942-8925. ✄

Photo by D. Elliott, courtesy of Applied Bio-nomics, Ltd.

spider mite controls

The All-Purpose
Predatory Mite

Neoseiulus = Amblyseius fallacis

(Nee-oh-SAY-yu-luss =AM-bleh-say-uss fah-LAY-shiss)

Description— "Mighty mites." That's what we think they should be called. These mites, like tiny spiders, eight legs and all, are voracious predators of **several pestiferous spider mite species**. We have personally had phenomenal results with their use. So have the vast majority of our contacts who've tried them.

N. fallacis are shipped as adults and immatures in vermiculite (see photo, below) or as adults and immatures, with a bonus of some eggs, on bean leaves (see photo, next page).

The vermiculite product is flowable and easy to distribute. The leaves, however, are also an excellent distribution medium, and seem to go a little further for some folks. Moreover, the predatory mites are more comfy during their travels on the leaves. These are the same leaves upon which they're reared. We feel the leaf product is slightly superior to the other for this reason.

N. fallacis can prevent *and* control, as said above, a number pest mites in a multitude of conditions. Some of the species they can impact include: **the two-spotted mite (*Tetranychus urticae*); the carmine red mite (*T. cinnabarinus*); a two-spotted mite relative (*T. evansi*); the European red mite (*Panonychus ulmi*); the citrus red mite (*Panonychus citri*); the southern red mite (*Oligony-***

N. fallacis provided in a vermiculite carrier.

Photo by M. Cherim

chus ilicis); the Brevipalpus citrus mites (*Brevipalpus californicus, phoenicis* and *obovatus*); the six-spotted mite (*Eotetranychus sexmaculatus*); the Texas citrus mite (*Eutetrannychus banksi*); the tumid spider mite (*T. tumidus*); the Pacific mite (*T. pacificus); and,* perhaps, the *Phalanopsis* mite (*Tenuipalus pacificus*). Moreover, these predators may offer some control of the privet mite (*B. obovatus*), cyclamen mites (*Phtyodromus =Steneotarsonemus pallidus*), broad mites (*Polyphagotarsenomus =Hemitarsonemus latus*) and tomato russet mites (*Aculops lycopersici*), and other species. These mites also consume pollen — they can live on it (see Benefits for more information).

Life-style— The tiny 0.5 mm. clear-white to clear-pink adult female mites lay eggs amongst spider mite concentrations and their webbing (which is produced by the two-spotted mite), if present. They can lay up to 100 of them! The eggs hatch into super small larvae which develop into nymphal forms before reaching adulthood. These, too, are fierce predators, consuming many spider mites eggs and young.

The life-span of these predators is roughly 8 days in their immature stages, then around 1 month as adults. The conditions for optimum performance will be between 50-80°F with a relative humidity of between 60-90%. But these are *optimum* conditions, and not necessarily a prerequisite of successful implementation. Please note, however, considerably cooler and warmer temperatures will hamper reproduction and development a certain degree (see Benefits).

Benefits— These mites will feed at temperatures as low as 35°F and as high as 100°F. However, at the low-end of the scale (<50°F), they won't reproduce, and at the high-end they need very humid conditions to work with any efficiency (see Advisories).

N. fallacis provided on bean leaves as a carrier. This choice is best for mites needing a soft ride to their job site. This product is also pyrethrin resistant.

Photo by M. Cherim

N. fallacis are very cost-effective — the most cost-effective. And at the recommended rates, they can be a very loud bang for the buck!

These predators can live off of pollen alone. (See Advisories.) This makes them an excellent preventive agent as well as a curative one. After all, the insectary where the leaf product is produced is still labeling them as a field mite preventive (see photo, facing page). This, however, is no longer really the case (see Usages).

The leaf product has also been specially adapted to tolerate pyrethrins and pyrethroids. These are the only beneficial species we discuss with this trait. Both strains, the pyrethrin resistant [PR] on leaves and the regular in ver-miculite, are compatible with a very large selection of pesticides, including some very powerful miticides (see the Biorational Substances Chart on page 3 for details).

Ah! Pollen. *N. fallacis* love it. And because of this, we love them. Having the ability to live at a site, consuming pollen, happy, *before* a pest mite rears its ugly head, is a real advantage. Most predatory mites would have to leave the area or die without prey. These predators *can* clean up an infestation. But, better yet, they can prevent one from occurring. (See Advisories and Rates.)

Lastly, one more benefit: these mites will overwinter and establish them-selves just about everywhere!

Drawbacks— As many of our contacts know, these mites are sometimes in limited supply during certain times of the year. This is because of seasonal demands being extremely high.

Excessive webbing, associated with *severe* infestations, can interfere with the performance of these mites.

Distribution of mites *within* the leaf carrier may be uneven. Concentrations may be found on one leaf, while there may be none on another.

Scouting— Un-infested and damage/webbing-free new growth is a good sign. So is empty webbing, in interiorscapes and some greenhouses, or missing webbing, outdoors (if it was there to begin with). For more on webbing, see Fig. 6, bottom of page.

Unless your scouting is really top-notch, you'll probably miss most of the predators present on the leaves. However, if you see some agile-looking mites running quickly across the leaf's undersurface, they are probably predators.

Be a WEBWIPER

Fig. 6

Here's a physical means of pest control that's worth looking into: web-wiping. 1) With a damp sponge, and maybe a little isopropyl alcohol, wipe away the webbing on the plant's leaves. 2) Throw the sponge in your washing machine. 3) You're done, that's it. Doing this, you'll destroy literally hundreds of adult mites, their young and many eggs. The webbing is mite-central-station. Just be careful not to pass the mites on to other plants. ✠

Advisories— Flowering, pollen producing plants are a big plus, since, as mentioned previously, these predators might stay on-site if there is a supplemental food source.

Preventive releases are very cost-effective and useful, especially when pollen and sheltering groundcover are present. Spring release are the obvious time that preventive releases can be made. However, under many circumstances, fall releases will also work well with this mite in the name of prevention. This is especially true on strawberries.

Low-growing crops like strawberries allow these mites to work all summer long, in the hottest of conditions. The reason, we feel, is the more humid microclimate close to the ground and under the canopy of the plants' leaves. To artificially recreate these conditions, try a little foliar misting. Most pernicious varieties of mites prefer hot, dry conditions. Consequentially, between the coolness and humidity increase registered when misting, you will be hampering the pests as well as helping the beneficials.

Usages— Where can't they be used would a more appropriate question. We've seen the successful implementation of these species in just about every conceivable situation. Especially in greenhouse and outdoor straw- and cane-berries, and interiorscape plantings and palms. These mites also work great on ornamentals, bedding plants, hedges, in trees, orchards, fields and more.

Rates—

```
PRVNT: 5-9 / YD. - *MTHLY - NEED
CTRL-L: 10-18 / YD. - TRI-WKLY - 2-3X
CTRL-M: 18-26 / YD. - BI-WKLY - 2-4X
CTRL-H: 26-32 / YD. - BI-WKLY - 2-4X
MAINT: 3-6 / YD. - MTHLY - INDEF
GRDN: 25-45% of rates above.
ACRE+: 5-30% of rates above.
COMMENTS: *Fall preventive releases should
be made into groundcover only one time.
```

Pricing information— The *N. fallacis* mites covered in this section can be obtained from The Green Spot by calling 603/942-8925. Detailed release instructions are provided with every order. Here is the current, industry-average pricing for 1998...

Item no. CNF2M ... 2k. N. fallacis in vermiculite ... 1 unit = $13.44

Item no. CNF10M ... 10k. N. fallacis in vermiculite ... 1 unit = $50.70

Item no. CNF10M-6 ... 10k. N. fallacis in verm. ... 6-11 units = $44.36 ea.

Item no. CNF10M-12 ... 10k. N. fallacis in verm. ... 12+ units = $39.92 ea.

Item no. CNFPR25C ... 2500 N. fallacis on leaves ... 1 unit = $32.62

Item no. CNFPR5M ... 5000 N. fallacis on leaves ... 1 unit = $45.50

Item no. CNFPR5M-12 ... 5000 N. fallacis on leaves ... 12+ units = $38.67 ea.

The Fast-Action
Predatory Mite

Phytoseiulus persimilis
(fii-toh-SAY-yu-lus per-SIM-ill-iss)

Description— "Killers." That's what they are. They can't go long without food. They must continually hunt and eat, hunt and eat, hunt and eat.

These mites, like tiny spiders, eight legs and all, are voracious predators of most of the ***Tetranychus*** species. We have personally had phenomenal results with their use. So have the vast majority of our contacts who've used them.

P. persimilis are shipped as adults and immatures in vermiculite (see photo, below) or as adults and immatures, with a bonus of some eggs, on bean leaves (see photo, next page).

The vermiculite product is flowable and easy to distribute. The leaves, however, are also an excellent distribution medium, and seem to go a little further for some folks. Moreover, the predatory mites are more comfy during their travels on the leaves. These are the same leaves upon which they're reared. We feel the leaf product is slightly superior to the other for this reason.

P. persimilis can control a number pest mites belonging to the genus ***Tetranychus***. Some of the species they can impact include: **the two-spotted mite (*T. urticae*), the carmine red mite (*T. cinnabarinus*), the tumid spider mite (*T. tumidus*), and the Pacific mite (*T. pacificus*).**

Life-style— The tiny 0.5 mm. hunter-orange female mites lay eggs amongst spider mite concentrations and their webbing (which is produced by the two-spotted mite), if present. They can lay up to 60 eggs! They hatch into minuscule larvae which develop into nymphal forms before

P. persimilis in a vermivulite carrier.

Photo by M. Cherim

reaching adulthood. These, too, are fierce predators, consuming many spider mites eggs and young.

The life-span of these predators is roughly 8 days in their immature stages, then around 36 days as adults. The conditions for optimum performance will be between 70-85°F (extended to 60-90°F) with a relative humidity of between 60-90%. But these are *optimum* conditions, and not necessarily a prerequisite of successful implementation. Please note, however, cooler temperatures will hamper reproduction and development a certain degree.

Benefits— These mites are gluttonous, there's no other way to look at it. And their gluttony is to your advantage. If you put *P. persimilis* in a suitable environment with nearly any crop to clean up two-spotted mites, that's exactly what they'll do. Very fast, very active, very thorough.

P. persimilis are very cost-effective — not the *most* cost-effective, but worthwhile anyway. And at the recommended rates, they can be a very loud bang for the buck! These predators are an excellent curative agent.

Drawbacks— If there are no spider mites suitable to the tastes of *P. persimilis* available, they'll quickly starve to death. Unlike *N. fallacis* which can support its needs with pollen, or *N. californicus* (no longer discussed in this manual) which can walk around as a wafer-thin disk endlessly eating nothing, *P. persimilis* must have fresh meat. Its preventive skills amount to none.

This predatory mite species will not do very well in cold climates (see *N. fallacis*).

P. persimilis on their natural, bean leaf carrier

Photo by M. Cherim

Scouting— Uninfested and damage/webbing-free new growth is a good sign. So is empty webbing, in interiorscapes and some greenhouses, or missing webbing, outdoors (if it was there to begin with).

Unless your scouting is really top-notch, you'll

This P. persimilis adult is orange in color and as long, fast legs which allow it to hunt very effectively for slower-moving mites like the two-spotted mites.

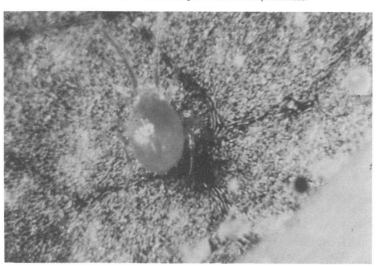

probably miss most of the predators present on the leaves. However, if you see some orange, long-legged agile-looking mites running quickly across the leaf's undersurface, they are probably these predators (see photo, above).

Advisories— Be certain of your pest species. *P. persimilis* are not as flexible as some other available species. These guys like *Tetranychus* species only. And not all of those species either.

Low-growing crops like strawberries allow these mites to work all summer long, in the hottest of conditions (this is true in California where these mites are often used in strawberries and other like-crops). The reason, we feel, is the more humid microclimate close to the ground and under the canopy of the plants' leaves. To artificially recreate these conditions, try a little foliar misting. Most pernicious varieties of mites prefer hot, dry conditions (see spider mites). Consequentially, between the coolness and humidity increase registered when misting, you will be hampering the pests as well as helping the beneficials.

The *P. persimilis* leaf product will often contain a little surprise for its users at no extra cost: *Feltiella acarisuga*, a predatory spider mite midge similar to *Aphidoletes aphidimyza* (see Aphid Controls). For more information regarding *F. acarisuga*, see Fig. 7, bottom of next page.

Usages— Where can't they be used would be a more appropriate question. We've seen the successful implementation of these species in just about

Photo by B. Costello, courtesy of Applied Bio-nomics, Ltd.

every conceivable situation, except tomatoes where they don't perform as well. If you have the appropriate species and number of mites, and if the conditions are right, control is virtually inevitable.

Rates—

```
PRVNT: N/A
CTRL-L: 7-11 / YD. - TRI-WKLY - 2-3X
CTRL-M: 10-18 / YD. - BI-WKLY - 2-4X
CTRL-H: 17-29 / YD. - BI-WKLY - 2-4X
MAINT: 4-9 / YD. - MTHLY - INDEF
GRDN: 45-65% of rates above.
ACRE+: 25-40% of rates above.
COMMENTS: For tomatoes and other
solanaceous plants, double rates above.
```

Pricing information—
The *P. persimilis* mites covered in this section can be obtained from The Green Spot by calling 603/942-8925. Detailed release instructions are provided with every order. Here is the current, industry-average pricing for 1998...

Item no. CPP5C ... 500 P. persimilis in vermiculite ... 1 unit = $10.40

Item no. CPP5C-6 ... 500 P. persimilis in vermiculite ... 6+ units = $8.32 ea.

Item no. CPP1M ... 1000 P. persimilis in vermiculite ... 1 unit = $13.52

Item no. CPP1M-6 ... 1000 P. persimilis/vermiultie ... 6-11 units = $11.83 ea.

Item no. CPP1M-12 ... 1000 P. persimilis/verm ... 12+ units = $10.82 ea.

Item no. CPP2M ... 2000 P. persimilis in vermiculite ... 1 unit = $24.96

Item no. CPP2M-6 ... 2000 P. persimilis/vermiultie ... 6-11 units = $22.46 ea.

Item no. CPP2M-12 ... 2000 P. persimilis/verm ... 12+ units = $19.98 ea.

Item no. CPPL15C ... 1500 P. persimilis on leaves ... 1 unit = $20.80

Item no. CPPL15C-4 ... 1500 P. persimilis on leaves ... 4-7 units = $19.24 ea.

Item no. CPPL15C-8 ... 1500 P. persimilis on leaves ... 8+ units = $17.68 ea.

FIG. 7

Feltiella acarisuga

These small, delicate midges are very similar to A. aphidimyza, the aphid midge, except for two distinct differences: the predacious larvae of these midges prefer spider mites to aphids, and they don't pupate in the soil (they do so on the surface of leaves) and thus have no compatibility difficulties with parasitic nematodes.

We offered these midges in 1996. They were supplied on cucumber leaves. It didn't work out though; they were too difficult to rear in an isolated culture. Now, however, you can get them for free sometimes (usually all winter long) when you order the P. persimilis product on leaves. 🞜

The Spider Mite
Destroyer

Stethorus punctillum
(Steth-OR-uss punk-TILL-umm)

Description— These tiny 1.5-2 mm. black beetles are relatively new to the commercial bio-control industry. However, naturally speaking, they've been around for a very long time and are fairly well documented; since as early as 1936. Needless to say — but said anyway — we're happy to be able to tell you about them and participate in their availability.

One thing which is a certainty: these beetles, as their name implies, have a pretty impressive track record for controlling **various spider mites**. We are not aware of alternate prey for these little guys, but, being beetles — which are fairly opportunistic by nature — they may impact other critters. There is no indication that *predatory spider mites* are at risk of being eaten, however. This is probably because predatory mites are very fast and tend not to hang out on one leaf for too long.

S. punctillum are very similar in appearance to *D. pusillus* (see whitefly controls), but tend to be a little hairy like the scale eating cousins, *R. lophanthae* (see scale controls).

These beetles have historically controlled the **two-spotted mite and the carmine red mite (*Tetranychus urticae* and *T. cinnabarinus*, respectively); the European red mite (*Panonychus ulmi*)**; and another species, *Metatetranychus ulmi*, of which its common name, if it has one, is unknown, but may be a synonym for the European red mite mentioned above. It would not be at all surprising to discover many other pestiferous mites which could, at least, be impacted by this aggressive little beetle.

S. punctillum are provided in small lots as pre-fed, pre-mated adults which, theoretically, are supposed to track down spider mite populations, begin feeding and reproducing, without too much delay.

Life-style— The minuscule adult female beetles lay their eggs amongst pest colonies, laying up to 8 eggs per day (in the lab, perhaps more in real life). The eggs hatch into greyish, alligator-shaped larvae similar to that of *R. lophanthae* (see scale controls), but smaller. These, too, are fierce predators, growing up to 2 mm. long and consuming a great number of pests.

The life-span of these predators is roughly 18 days in their immature stages, then 4-5 weeks as adults. The conditions for optimum performance will be between 69-90°F with a relative humidity of around 60%. But these are *optimum* conditions, and not necessarily a prerequisite of successful implementation. Please note, however, cooler temperatures (<59°F) will hamper reproduction and development a certain degree.

Benefits— These predators should provide outstanding results in the short- and long-term. Our own trials, thus far, have yielded favorable results in outdoor orchard applications. Indoor applications, including greenhouse, have been documented previously by others.

They can fly. With exception to *Feltiella acarisuga* (see fig. 7, page 100), there are no other commercially available spider mite predators with this special ability (unless you count predatory mites on a windy day). Flight allows this beetle to enter the canopy of trees, without the aid of hand placement, much more quickly.

Another significant benefit to using these beetles is the way they're supplied: pre-fed, pre-mated. Oviposition or egg-laying can potentially occur very soon after release.

Drawbacks— We have little practical experience with these predators. However, past and present research is somewhat promising. We strongly suggest greenhouse growers, interiorscapers and orchardist try this species.

These beetles *may* be a bit on the pricy side. This will probably be the case until production peaks. This hinges on steady demand. We suspect that the results may far outweigh the price issue.

There are no firm application rates as of yet. The rates shown for these beetles currently are more or less an educated guess. We need your help to let us know if our prescribed rates are effective.

Scouting— Adult beetle presence, larval presence, reduction of pest numbers and new, clean plant growth. These are all signs that spider mite destroyers are hard at work.

The white eggs of this beetle can sometimes be found scattered along the host plant's leaves, top and bottom. Since many beetle eggs have a distinctive football shape, and are usually laid in rafts (clusters and/or rows). These eggs, found scattered, may be easier to spot.

The scout may also wish to search for the beetles' pupae — greyish balls — in the lee of the veins on the leaf undersides.

We have nothing more to offer at this point in time. And this, unfortunately, includes photographs and illustrations.

Advisories— Aside from misting the site with water before releasing and doing so in the evening (sometimes not necessary in interiors), there are other things you can do to ensure the maximum number of beetles perform. Flowering, pollen producing plants *may* a big plus, but still needs confirmation.

Honeydew is another possible supplemental food source, but it is unlikely a significant one.

You may want to try releasing these beetles into trees the way we are suggesting to try for Crypts: Find a long bamboo or wooden pole with an approximate diameter of three-sixteenths to one-quarter-inch; insert one end of the pole through the pre-manufactured hole in the bottle's lid and into the bottle of beetles, being careful not to injure any of them; the beetles will climb up the pole to the other end [if it's not too hot, in which case they may fly off the pole (but should still make it to the foliage — eventually)], which should be tucked into the tree's foliage.

Yellow sticky traps used in greenhouses *may* present a problem with these beetles. We are unsure of this and ask that you exercise cautious observation when the traps are in use. If predator numbers on the traps get too high, remove the traps or use them for only two or three days per week.

Usages— We're not certain of every possible use. We suggest trying them in any environment where pest mites are present. And please let us know how they do. This especially true of tomato growers who are eager for another form of mite control.

Rates—

```
PRVNT: 1-2 / YD. - MTHLY - NEED
CTRL-L: 3-4 / YD. - TRI-WKLY - 2-3X
CTRL-M: 4-5 / YD. - BI-WKLY - 2-4X
CTRL-H: 5-6 / YD. - BI-WKLY - 2-4X
MAINT: 1-3 / YD. - QTRLY - INDEF
GRDN: 45-65% of rates above.
ACRE+: 25-40% of rates above.
COMMENTS: For tomatoes and other
solanaceous plants, use 125% of rates.
```

Pricing information—

The *S. punctillum* beetles covered in this section can be obtained from The Green Spot by calling 603/942-8925. Detailed release instructions are provided with every order. Here is the current, industry-average pricing for 1998...

Item no. CSP1C ... 100 S. punctillum adults ... 1 unit = $42.00

Item no. CSP1C-4 ... 100 S. punctillum adults ... 4-7 units = $37.80 ea.

Item no. CSP1C-8 ... 100 S. punctillum adults ... 8 + units = $34.96 ea.

thrips

Order: Thysanoptera   Sub Orders: Terebrantia & Tubulifera

What a pain in the [expletive] these tiny pests can be. They investigate plant parts so thoroughly that, in many cases, they cannot be reached by conventional spray techniques. Some growers spray during the high-noon-hots in their greenhouses because the thrips seem to be more active and obvious during that time (see photo, below). Others use electrostatic spray equipment — which electrically charges the spray particles so they will adhere to every exposed surface. However, none of these techniques provide the kind of kill growers want.

This thrips (shown on the leaf's midrib) is surrounded by aphid mummies... obviously the aphid bio-control is working (see aphid controls).

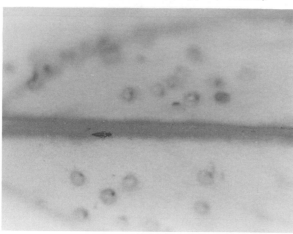

You may think that the paragraph above is going to lead you to a sentence or statement which boasts how much better than chemicals the predators and parasite used to control thrips are. Well, that's not the case. Sure they can be, but chemicals have their moments of glory too. The point is: thrips are tough as nails and no matter what means are taken to effect a change in their numbers, don't go soft, and don't lighten up, even when it appears things might be clean.

Thrips must be from hell. They should not be given an inch. They may take a mile. They can cause extensive plant damage (like that shown in the photo, facing page, top) and transmit two economically important diseases: impatiens necrotic spot virus [INSV] and the tomato spotted wilt virus [TSWV].

Thrips are very small, cigar-shaped insects (see photo, facing page, bottom). Their body design allows them access to their favorite plant parts: the buds and flowers.

Photo by M. Cherim

Thrips have rasping mouth-parts. They first abrade the plant's cells, rupturing them, then they take up the sap — or blood, if you will.

Thrips attacking the flower of a cucumber plant can cause extensive damage like the aborted cucumber in this photo.

The scout can find thrips with relative ease, despite their small size. First of all, if the infestation is high, damage will be readily visible as streaking or silvering of some of the leaves, speckling may also be a sign. Second of all, through the use of yellow, blue or *hot pink sticky traps, or by shaking plant parts — especially flowers — over a plain white sheet of paper, the thrips can easily caught and counted.

Should you be concerned? After scouting your crops for immediate problems, search areas surrounding your crops — outside the greenhouse, across the street, in the field beyond that. Feed corn behind your greenhouse might cause trouble. Should you be concerned? Well, yes, if you have susceptible crops and find thrips anywhere near them.

Thrips adult, with its feather-like wings folded back over body. These pests are rarely seen any other way.

If you find yourself in this last category, perhaps the most important thing you can do is to scout like a pro. Moreover, if the situation is bad enough — or potentially so — screening for thrips exclusion

*A 1996 Cornell University Cooperative Extension study determined that "hot pink" is the best sticky trap color for trapping and monitoring thrips (over blue, that is). GMPro magazine — the trade publication from which this information was obtained — (July 1997), gave anyone using blue traps a "thumbs down." With all this peer pressure, we think that you should use hot pink traps — but can someone please tell us who manufactures these?

Photo, top, by L. Gilkeson. Photo, bottom, by M. Herbut. Both courtesy of Applied Bio-nomics, Ltd.

These thrips' larvae will soon drop to the ground to pupate.

may be well worth the expenditure. Another way to help curb the development of thrips in your greenhouse or interior is to set a concrete slab floor. Most troublesome thrips' larvae (see photo, below) drop from the plant to pupate in the soil, media, gravel, weed cloth, etc. Sealing the floor with concrete will help stop this from occurring. Unfortunately, they can still drop in the pot — but every little bit helps. Please note: despite the comments above, sometimes the more human intervention plays a role, the more things get out of whack. Our advice, do only what's absolutely necessary (as dictated by Ma Nature).

For more detailed information and recommendations concerning thrips and ways to effectively control them, you are invited to call The Green Spot, Ltd. at 603/942-8925. ▨

Photo by M. Herbut, courtesy of Applied Bio-nomics, Ltd.

thrips controls

The Soil-Dwelling Mite

Hypoaspis miles
(Hi-poh-A-spiss mylz)

Description— If at gunpoint we had to choose a favorite organism, *Hypoaspis miles* might be what we'd answer. Why? The reason is the organism's use and potential — it's fantastic. Not only are these mites predators of **thrips' prepupal and pupal stages**, they are very effective **fungus gnat** predators as well.

Now for the next question: If asked if there is one organism which should be employed in nearly every greenhouse or interiorscape, worldwide, what would we say? *H. miles*. And we're not exaggerating.

H. miles are shipped as adults, immatures and eggs (the latter are not part of the guaranteed count) in half-liter- and liter-size shaker canisters

These are the one-liter H. miles shaker bottles.

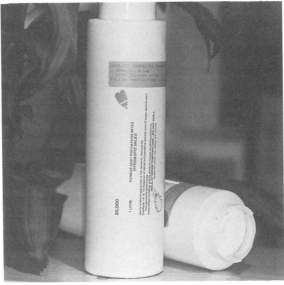

(see photo, inset) filled with a loose vermiculite carrier and a percentage of clean peat. In this form, *H. miles* are very easy to distribute in the crop.

H. miles can help prevent thrips from becoming intolerable. They can also be instrumental in assisting another predator with the timely and efficient control of an outbreak. Some of the species they can devour are listed under the thrips predator, *N. cucumeris* (discussed later in this section). Moreover,

Photo by M. Cherim

by themselves, they can control all the fungus gnat species listed in Fig. 4, on page 82. And if that's not enough, these little carnivores (meat eaters) / detritivores (recyclers) also eat **springtails (species in the insect order Collembola)**. Currently under investigation is the **potential for these mites to impact root mealybugs (*Rhizoecus arabicus* and other *Pseudococcidae* family members), sow bugs and pill bugs [a.k.a. rolly-pollies (*Portcellio* spp. and *Armadillidium vulgare*, respectively)], and even a fungus-eating mite (sp. unk.)** found in some Shiitake mushroom production houses. The

Here's a H. miles adult — out of its natural environment for the sake of contrasting photography.

investigation is showing the most promise where the impact of the mushroom mite is concerned (in addition to great mushroom fly control). As for the rest, it's still too early to tell.

Life-style—
These robust 0.8 mm. richly-colored tan to brown mites (see photo, left) live, eat and reproduce in the soil or soilless medium in containers and planters and, on the walkways (see photo, facing page) and floors of greenhouses (or on the Shiitake logs in the mushroom scenario). The *H. miles* females' eggs — of which there are many — hatch into super small larvae which develop into tiny dark brown, almost black, nymphal forms before reaching adulthood. These, too, are fierce predators, consuming many pests, mostly the eggs and smaller larvae (first and second instar) of fungus gnats and the pupal stages of thrips.

The life-span of these predators is about 13 days from egg to ever-after. But they reproduce profusely in what little time they have. The conditions for optimum performance will be between 60-72°F, and we're talking about soil temperatures, with a relative humidity equal to that found in a friable, slightly moist medium, compost or soil. But these are *optimum* conditions, and not necessarily a prerequisite of successful implementation. Please note,

Photo by D. Gillespie, courtesy of Applied Bio-nomics, Ltd.

however, considerably cooler temperatures will hamper reproduction and development a certain degree.

Benefits— *H. miles* are very cost-effective. A little bit goes a long way and they can last such a long time.

These predators, like *N. cucumeris*, are supplied with some mold mites (*Tyrophagus putrescentiae*). These mold mites are merely a *non-sustainable* food source (but only after the contents are distributed) for the predatory mites while they're in transit and in storage. That's right, *storage*, you can give a half-liter bottle to each interiorscape tech; the bottles store for two weeks — easily. The "Hypos" just go about their business, feeding on the mold mites and their own dead and miscellaneous stuff. Fair humidity in storage is helpful, and the techs can't leave the bottles in their cars, etc. Other than that, "have Hypos will travel!"

These predators are compatible with many —cides; a real plus in an IPM program (see the Biorational Substances Chart on page 6 for details). Part of the reason for this is simple: these mites are protected in the soil during spraying. They *do* wander at night, though, so caution is still advisable. This wandering, however, has its advantages: the mites can establish themselves in the *entire* area usually.

Here a greenhouse employee wisely distributes H. miles in any nook and cranny capable of supporting thrips and fungus gnat development.

Drawbacks— You get a lot, but you have to use a lot. Good thing they're inexpensive.

These mites like humid conditions. Therefore, growers of some herbs, most cacti and some succulents may not have suitable soil conditions for these predators. They probably won't have fungus gnats either. And as far as thrips...? Perhaps.

Photo by L. Gilkeson, courtesy of Applied Bio-nomics, Ltd.

H. miles can reduce thrips populations by only about 30%, but this isn't necessarily the fault of the mites (see Advisories).

Scouting— Using the paper method (discussed in greater detail later in this section, see *N. cucumeris'* Scouting) and yellow, blue and hot pink sticky traps to detect thrips is a sure way to progressively determine your predators' efficacy. Using yellow sticky traps (lying sticky-side-up at surface level) and potato disks (see Fig. 4, page 83) to monitor fungus gnat levels will do the same.

Unless your scouting is really top-notch, you'll probably miss most of the predators present in the soil or medium. But this doesn't have to be the case. First look at the mites in the container. (They're sometimes under the lid, if they're not readily visible, try blowing gently into the bottle. More often than not this will bring them up from the terrifying vermiculite depths.) This way you may be better able to identify the mites later. Now, in the release site (this works best in other than small containerized plantings) place stones, small pieces of bark, etc. around the site. Then, during your weekly scouting rounds, you can flip over the "little shelters" to look for your beneficial mites. But it's not unheard-of to find these tough survivors up to a year after a release. This isn't always the case. It happens mostly in large planters and greenhouse floors (see Advisories).

Advisories— For fungus gnat control, 6" pots and up are probably the best bet for containerized plant applications. If they're smaller than that, we normally recommend parasitic nematodes (see soil pest controls).

For thrips control, however, it is best to use the mites in nearly every application — but not alone. The thrips' in-ground stages make up only a small percentage of their overall life-span. This is why the *H. miles* can only handle about 30% of the overall population. An above-ground predator, like one of the other organisms in this section, really should be used. However, let it be known that the impact of *H. miles* on this extremely pernicious pest is noticeable at worst. 30% *is* 30%!

Do not rely entirely on the long-term establishment of *H. miles*. This happens, almost always to a *certain* extent, but not necessarily to such a degree that pest control can be maintained (see Rates). Multiple and regular releases are, like it is for our other bio-controls, recommended. One rule-of-thumb to consider: the bigger the container or release site the longer the interval between releases can potentially be (and thus fewer releases); and the opposite is true for smaller containers or growing areas.

Usages— Greenhouses/interiors are where these mites are most often used.

However, we've heard excellent reports from outdoor nurserymen who've had outstanding results.

Rates—

PRVNT: 70-90 / YD. - *MTHLY - NEED	
CTRL-L: 90-130 / YD. - TRI-WKLY - 2-3X	
CTRL-M: 130-190 / YD. - BI-WKLY - 2-4X	
CTRL-H: 190-270 / YD. - BI-WKLY - 3-5X	
MAINT: 80-100 / YD. - *MTHLY - INDEF	
GRDN: 40-70% of rates above.	
ACRE+: 30-60% of rates above.	
COMMENTS: For FUNGUS GNAT control use 50-75% of rates above. *Interval depends upon site-size.	

Pricing information—
The *H. miles* mites covered in this section can be obtained from The Green Spot by calling 603/942-8925. Detailed release instructions are provided with every order. Here is the current, industry-average pricing for 1998...

Item no. CHM10M ... 10k H. miles (half-liter) ... 1 unit = $16.90

Item no. CHM10M-8 ... 10k H. miles (one-liter) ... 8-11 units = $15.63 ea.

Item no. CHM10M-12 ... 10k H. miles (one-liter) ... 12+ units = $14.46 ea.

Item no. CHM20M ... 20k H. miles (one-liter) ... 1 unit = $22.88

Item no. CHM20M-8 ... 20k H. miles (one-liter) ... 8-11 units = $21.16 ea.

Item no. CHM20M-12 ... 20k H. miles (one-liter) ... 12+ units = $19.58 ea.

The Dark Mite

Iphiseius = Amblyseius degenerans
(iff-ih-SAY-uss =AM-bleh-say-uss dee-GEN-ehr-ranz)

Description— Persistent little predators, that's what they are — we think. The jury's still out on this one. Unless you're a pepper grower, or have a system capable of supporting long-term growth of this species, then we *know* these predators are top-notch. We also know, that regardless of crop, they love **thrips, spider mites** and pollen (see Drawbacks and Advisories).

I. degenerans are shipped as adults, with some immatures and eggs as well, at times (the latter are not part of the guaranteed count). They're provided in a couple of different size plastic leaf trays, in which there are some bean leaves with a touch of bee pollen for the traveling mites (see photo, next page). Other carriers, such as vermiculite, are currently being looked into.

Ahhh, 500 of the finest Dark Mites, resting in a bed of the freshest bean leaves, with just a hint of *bee pollen. Presented in a fine faux-crystal dish.
*(To prevent the mites from becoming mortally challenged in transit).

I. degenerans can potentially prevent thrips from becoming intolerable. And in the right conditions, they can establish themselves and provide ongoing control through continued preventive activities. Some of the species they can devour include all of the **species controlled by *N. cucumeris*** (see next, this section). This *possibly* includes the **mite species** as well.

Other mites can also fall prey to *I. degenerans*, some of the *Tetranychus* spp. for example (see spider mite controls). We do NOT recommend using these mites for other pests at this time, unless it is for the sake of experimentation. We believe control of other pests might not be obtainable with typical dark mite releases. Coincidental cross-predation should be regarded as a bonus of the application and nothing more. Conduct your own trials if you wish. Get back to us please; let us know how well they work; tell us what you saw. Like most of the recommendations in this manual, we learn these truths from our clienteles' experiences. (Others are based on producer recommendations, our own research trials, etc.)

Life-style— The slick-looking, tiny 0.7 mm. dark-brown (they appear to be black) adult female mites lay eggs amongst thrips' concentrations and close to pollen sources. The eggs hatch into super small larvae which develop into nymphal forms before reaching adulthood. These, too, are effective predators, consuming many immature thrips.

The life-span of these predators is about 9 days in their immature stages, then around 30 days as adults. The conditions for optimum performance will be between 65-85°F with a relative humidity of between 60-85%. But these are *optimum* conditions, and not necessarily a prerequisite of successful implementation. Please note, however, cooler temperatures will hamper reproduction and development a certain degree.

Photo by M. Cherim

Benefits— These predators enjoy snacking on available pollen (see Drawbacks and Advisories.) This helps them be an efficient long-term preventive agent as well as a potential curative one, as already discussed.

These mites are a bit on the pricy side. However, because of their ability to establish, they can be *extremely* economical. This is currently the case for sweet pepper production. Pepper growers are having extraordinary success.

Drawbacks— These predators may actually like pollen over pests in some cases. This should be considered when applying this species. Close observation is definitely warranted (see Advisories). We are still investigating this possibility.

We don't know too much about these beautiful mites. For one thing, only a small segment of our contacts have tried them. And the results have been there, but not definitively.

Historically, supply has been shaky. This was a drawback. Perhaps it still may be at times. The progress with improved production techniques thus far, though, is promising. Availability may not be a problem this year.

Extensive webbing from spider mites, if present, may hamper the efficiency of the dark mites. We would still use them in a dual-pest situation like this, though. Especially since they will eat the mites, too. But, first, be a webwiper (see Fig. 6, page 95).

Harsh climates cannot support this species. With exception to indoor applications, northern use outdoors is probably not an economical option.

Scouting— Finding *I. degenerans* on the plants, because of their dark, contrasting color, is relatively easy for the amateur scout.

But, nevertheless, it can take some practice, the mites *are* pretty small. In addition to looking for predators, we suggest simply monitoring the thrips levels. This is still the easier option.

Monitoring thrips levels can be done two different ways: 1) use blue, yellow or hot pink sticky traps to capture adults which can be counted weekly (we discuss the traps in our scouting section) and/or; 2) use the "paper method" by gently shaking the plant over a white piece of paper on which they'll show up nicely. Try the latter on a corn plant this summer, tap out the tassels if you really want to see some thrips.

Advisories— Flowering, pollen producing plants may be a plus, *maybe*. It is suspected that the mites might consume pollen more readily than pests. On this we are not yet totally clear. Each datum received will be passed on to our callers; as we learn, so will you. Until we have the answers, we'll encourage you to provide pollen or a source thereof, to these mites. Just avoid providing too much; let's try keeping the mites lean and mean.

There may be an *I. degenerans* compatibility issue when used with *N. cucumeris* (see, this section). Used together, this pair is under suspicion: the *I. degenerans* is suspected of being able to run *N. cucumeris* out of town.

Usages— Greenhouses (especially on greenhouse peppers), interiors, and the like, are where these mites are most often used. They can also be used in some nurseries for outdoor thrips control (as a trial).

Rates—

> PRVNT: 2-4 / YD. - MTHLY - NEED
> CTRL-L: 5-9 / YD. - TRI-WKLY - 2-4X
> CTRL-M: 10-14 / YD. - BI-WKLY - 2-5X
> CTRL-H: N/A
> MAINT: 2-3 / YD. - MTHLY - INDEF
> GRDN: 50-60% of rates above - TRIAL.
> ACRE+: 30-40% of rates above - TRIAL.
> COMMENTS: Winter preventive releases in peppers should be around 25-50% of rates above.

Pricing information—
The *I. degenerans* mites covered in this section can be obtained from The Green Spot by calling 603/942-8925. Detailed release instructions are provided with every order. Here is the current, industry-average pricing for 1998...

Item no. CID1C ... 100 I. degenerans on leaves ... 1 unit = $22.50

Item no. CID1C-6 ... 100 I. degenerans on leaves ... 6+ units = $19.12 ea.

Item no. CID5C ... 500 I. degenerans on leaves ... 1 unit = $63.00

Item no. CID5C-4 ... 500 I. degenerans on leaves ... 4-7 units = $53.55 ea.

Item no. CID5C-8 ... 500 I. degenerans on leaves ... 8+ units = $48.19 ea.

The Thrips

Predatory Mite

Neoseiulus = *Amblyseius cucumeris*

(nee-oh-SAY-yu-luss =AM-bleh-say-uss KU-KU-mer-iss)

Description— Like most of the *Neoseiulus* species, *N. cucumeris* are tough, flexible predators. *These* happen to prefer thrips, mostly the immature stages.

N. cucumeris are shipped as adults, immatures and eggs (the latter are not part of the guaranteed count) in a loose bran-flake carrier. With this medium, these predators are supplied three different ways: a bulk product for fast distribution and consumption (see photo, below); a pre-punched packet product for slow preventive releases over an 8 week period; and a packet which is not pre-punched that must be torn open and immediately distributed. The latter is designed for extremely small scale use.

N. cucumeris, in either form, can prevent thrips from becoming intolerable. The bulk unit and small quick release packet can provide control. Some of the species they can devour include: the **western flower thrips (*Frankliniella occidentalis*); the flower thrips or eastern flower thrips (*F. tritici*); the onion thrips (*Thrips tabaci*); the greenhouse thrips (*Heliothrips haemorrhoidalis*); and possibly the melon thrips (*Thrips palmi*), too.** Other pests which can be impacted by these predators include **cyclamen mites (*Phtyodromus =Steneotarsonemus pallidus*), broad mites (*Polyphagotarsenomus =Hemitarsonemus latus*) and, to a slight degree, tomato russet mite (*Aculops lycopersici*).** And, belonging to the genus *Neoseiulus =Amblyseius*, these predators may eat

This liter-size shaker canister contian 50,000 mites in bran flakes.

Photo by D. Simser

other pests as well.

Life-style— The tiny 0.5 mm. clear-white adult female (see photo, below) mites lay eggs amongst thrips concentrations. They can lay up to 35 eggs. The eggs hatch into super small larvae which develop into nymphal forms before reaching adulthood. These, too, are fierce predators, consuming many immature thrips.

The life-span of these predators is about 10 days in their immature stages, then around 30 days as adults. The conditions for optimum performance will be between 66-80°F with a relative humidity of between 65-72%. But these are *optimum* conditions, and not necessarily a prerequisite of successful implementation. Please note, however, considerably cooler temperatures will hamper reproduction and development a certain degree.

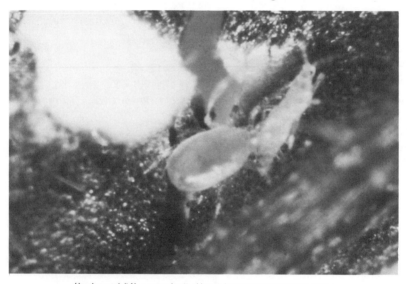

Here's an adult N. cucumeris attacking an immature western flower thrips.

Benefits— *N. cucumeris* are very cost-effective.

These predators enjoy snacking on available pollen (see Advisories.) This helps them be an efficient long-term preventive agent as well as a curative one. And, when supplied in slow-release packets, they are even more outstanding. The packets, which normally last up to eight weeks, contain a food source for the predators: a bran mite or, more properly, the mold mite (*Tyrophagus putrescentiae*). This mold mite is merely a sustainable food source for the predatory mites while they're in transit, and longer in the packet system (see Advisories for more).

Photo by M. Herbut, courtesy of Applied Bio-nomics, Ltd.

These mites are very, very economical. So economical, in fact, some growers are using *N. cucumeris* as they would a pesticide: the infamous repeated dousing technique.

Speaking of pesticides, these predators are compatible with many; a real plus in an IPM program (see the Biorational Substances Chart on page 6).

Lastly, commercially produced *N. cucumeris* is mostly nondiapausing strain. They can, therefore, be used year 'round.

Drawbacks— You get a lot, but you have to use a lot. Good thing they're inexpensive.

The packets are great, though more expensive, they do the preventive job assigned to them. The problem is when growers want to use them for control in an environment where the loose material is impractical to apply (see photo, right): certain hydroponic systems, for example where the bran carrier may fall into a trough. The drawback is they're not designed to work that way. Slow release in the case of *these* packets means that the resident generation usually doesn't venture out of the bag. Their offspring, however, begin the process of prevention as they start exploring the world outside.

N. cucumeris do very well on cucumbers; and they're easy to apply to the broad, hairy leaves. The material "sticks" well.

For hydroponic growers, or others whose environment does not allow the distribution of the loose material, we recommend the use of distribution boxes. These boxes are useful for nearly any of the "loose" products, with exception, perhaps, to green lacewings (see aphid controls). They may be too cannibalistic.

Humid environments can sometimes be a drawback with the use of the packets (see Advisories).

Photo by L. Gilkeson, courtesy of Applied Bio-nomics, Ltd.

Scouting— Finding *N. cucumeris* on the plants is difficult. To the beginner we suggest simply monitoring the thrips levels. This is much easier than looking for predators all day.

Monitoring thrips levels can be done two different ways: 1) use blue, yellow or hot pink sticky traps to capture adults which can be counted weekly (we discuss the traps in the scouting section) and/or; 2) use the "paper method" by gently shaking the plant over a white piece of paper on which they'll show up nicely. Try the latter on a corn plant this summer, tap out the tassels if you really want to see some thrips.

Unless your scouting is really top-notch, you'll probably miss most of the predators present on the leaves. However, if you see some agile-looking mites running a zigzag quickly across the leaf's undersurface, they are probably predators.

Advisories— Flowering, pollen producing plants are a big plus.

Caution, we're told, must be taken if these mites are to be used with orchids. It is believed the mold mites (their food) can wreak havoc on orchids (they don't touch other crops, except bran, etc.). It is probably best to avoid using them on orchids altogether — just to be on the safe side.

A Washington State contact had poor results with the packet product. Conditions in that client's area are extremely humid and moist, famous for its rainfall. The packets began developing a *growth* of some sort about two weeks after they were put out. Consequently, we learned they have a limitation, and thus humid conditions (90% and up) should probably be treated another way.

That same grower does report good results with the loose bulk bran-product.

There may be an *N. cucumeris* compatibility issue when used with *Phytoseiulus persimilis* (see spider mite controls) and *Iphiseius degenerans* (this section). Used together, these pairs are under suspicion: the *N. cucumeris* may eat the other predators' eggs. And *I. degenerans* is suspected of being able to run *N. cucumeris* out of town.

Before getting some loose bran product, determine how it's going to be applied so that you can prepare. Beside shaking out the bulk bran onto the leaves, you may wish to apply it with a broadcast granule or whirlybird type spreader.

Usages— Greenhouses/interiors are where these mites are most often used.

We've been told that these mites cannot be used outdoors. However, looking at this logically, all insects and mites can survive and prosper *somewhere* outside. I mean, *really*! (Anyway, one of our more renegade-type contacts tried them outdoors, and they worked.) Shhh! Don't tell the nay-sayers but we offer some trial outdoor rates.

Rates—

PRVNT: 30-90 / YD. - *TRI-WKLY - NEED
CTRL-L: 90-150 / YD. - WKLY - 3-5X
CTRL-M: 150-210 / YD. - WKLY - 3-5X
CTRL-H: 210-270 / YD. - WKLY - 4-6X
MAINT: 60-120 / YD. - *MTHLY - INDEF
GRDN: 60-80% of rates above - TRIAL.
ACRE+: 50-75% of rates above - TRIAL.
COMMENTS: *The PRVNT and MAINT releases may be made with packets, repeated every 8 wks.

Pricing information—
The *N. cucumeris* mites covered in this section can be obtained from The Green Spot by calling 603/942-8925. Detailed release instructions are provided with every order. Here is the current, industry-average pricing for 1998...

Item no. CNC1M ... 1000 N. cucumeris fast-pkt ... 1 unit = $4.80

Item no. CNC1M-12 ... 1000 N. cucumeris fast-pkt ... 12-23 units = $3.84 ea.

Item no. CNC1M-24 ... 1000 N. cucumeris fast-pkt ... 24+ units = $3.07 ea.

Item no. CNC50M ... 50k N. cucumeris in bulk-bran ... 1 unit = $25.20

Item no. CNC50M-8 ... 50k N. cucumeris bulk-bran ... 8-11 units = $23.31 ea.

Item no. CNC50M-12 ... 50k N.c. bulk-bran ... 12+ units = $21.56 ea.

Item no. CNCP1C ... 100 N.c. slow-pkts (300/pkt) ... 1 unit = $50.70

Item no. CNCP1C-4 ... 100 N.c. slow-pkts ... 4-7 units = $45.63 ea.

Item no. CNCP1C-8 ... 100 N.c. slow-pkts ... 8+ units = $41.52 ea.

The Insidious
Flower Bug

Orius insidiosus
(or-ee-USS inn-sid-EE-oh-suss)

Description— By far, the most effective thrips predator is the true bug, *Orius insidiosus*. These little dynamos are capable of cleaning up large infestations of pests. And, in the right conditions, maintaining the newly cleaned site.

In addition to **thrips (all the species listed in this section, see *N. cucumeris*' Description), these predators like small larvae (caterpillars, etc.), an array of insect eggs, and a host of other organisms including aphids, mites, whiteflies, etc.**

O. insidiosus acts as a "general" predator. This is common among "true bugs." True bugs tend to be thorough at whatever they do. In the case of these predators, they can thoroughly clean house.

With few exceptions, we do NOT recommend using these bugs for the control of other pests, unless it is for the sake of experimentation. We believe control of other pests might not be satisfactorily obtainable with typical *O. insidiosus* releases. Coincidental cross-predation should be regarded as a bonus of the application and nothing more.

Here's an adult insidious flower bug, shot from above. A side view would show us how flat these bugs are — which helps them enter flowers and buds.

Moreover, these predators should still be used with *Hypoaspis miles* as a supplemental thrips control. After all, thrips deserve only the very best.

Photo by M. Badgley, courtesy of Buena Biosystems, Inc.

Insidious flower bugs are supplied as adults (see photo, facing page) with some nymphs present, or as nymphs alone (see photo, below). Both the adults and nymphs behave as predators, though the nymphs cannot fly as can the adults. They are supplied in a bottle (see photo, bottom) containing a mixture of mostly buckwheat hulls with a little vermiculite and some nonviable eggs of the moth *Ephestia kuehniella* as a temporary food source.

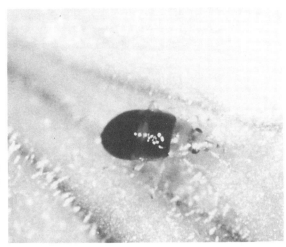

Here's the O. insidiosus nymph. It can't fly, but it's still mean as can be.

Life-style— The midsize 2.5 mm. adult female bugs lay eggs in plant tissue — up to 100 of them! The nymphs which hatch may taste the plant, but will cause no noticeable damage. By the time the nymphs reach their second instar (growth stage), they are ready for meat.

Bottle contains 500 bugs, buckwheat hulls, vermiculite and moth eggs as a temporary food source.

O. insidiosus, like other predatory true bugs, attack their prey with a needle-sharp proboscis through which pre-digestive enzymes are slowly exchanged for the bodily fluids of the prey. The proboscis is a sharp, straw-like mouthpart. It is certainly a weapon to be respected.

The life-span of these predators is about 2

Photo, top, by M. Herbut, courtesy of Applied Bio-nomics, Ltd.
Photo, bottom, by D. Simser

weeks in their immature stages, then 3-4 weeks as adults. The conditions for optimum performance will be between 62-84°F with a relative humidity of 60-85%. But these are *optimum* conditions, and not necessarily a prerequisite of successful implementation. Please note, however, cooler temperatures will hamper reproduction and development a certain degree.

Benefits— Being a thrips vacuum cleaner is not a bad trait of these predators. They are active searchers and flyers (adults only) and can really kick some bug butt. Just looking at them and seeing their fierceness is enough to instill confidence.

O. insidiosus can be establish themselves at a site if their needs are addressed. This *can* be economical. Simple amenities such as prey and pollen, upon which they can also feed, need to be available (see Advisories).

Drawbacks— Another drawback, but one hardly worth mentioning, is that these bugs, like the *Chrysoperla* spp. and *Deraeocoris brevis* (see aphid controls) can deliver a painful little bite (to people and each other). This is not to scare you. It's insignificant compared the benefits. And never have we had negative client feedback regarding this slight drawback.

Like *A. aphidimyza* and *D. brevis*, *O. insidiosus* undergo diapause (a quiescent state, hibernation) when the photoperiod is less than 10 hours (D) so long as temperatures are maintained at 73°F. If temperatures are going to be less, photoperiod should be held at 14 hours (D).

Pollen, thrips, etc. These are all good reasons for the insidious flower bug to take a moment to stop and smell the flowers.

Scouting— We suggest simply monitoring the thrips levels. This is still the easiest, most effective option.

Monitoring thrips levels can be done two different ways: blue, yellow or hot pink sticky traps and/or "the paper method" (see the preceding bio-

Photo by D. Simser

Adults *can* be found if the scout exercises perseverance. Try looking in the flowers (see photo, facing page) when the they may be seen eating thrips or pollen (see Advisories).

Advisories— To counteract the natural urge for these predators to undergo diapause you must 1) keep the temperature above 73°F, 2) keep temperatures above 50°F, but provide supplemental lighting during the appropriate time of year. A 60 watt bulb for every 60 foot radius *may* do the trick. We know that blue spectrum lighting (cool-white florescent) will do the trick. These predators, like *D. brevis*, are a little more difficult to fool than are *A. aphidimyza*. Bear in mind, also, most organisms, regardless of nature, will normally slow down a degree in the winter months.

Flowering, pollen producing plants are a big plus. The use of trap- and/or banker-crops is highly recommended.

Usages— *O. insidiosus* are useful in greenhouses, fields, interiorscapes, orchards and gardens. We recommend the successful implementation of these species in nearly every conceivable situation.

Rates—

```
PRVNT: N/A
CTRL-L: 2-3 / YD. - TRI-WKLY - 2-3X
CTRL-M: 4-5 / YD. - BI-WKLY - 2-4X
CTRL-H: 6-7 / YD. - WKLY - 2-4X
MAINT: 1-2 / YD. - MTHLY - INDEF
GRDN: 30-45% of rates above.
ACRE+: 5-20% of rates above.
COMMENTS: Use with Hypoaspis miles for
mid- to long-term kill.
```

Pricing information— The *O. insidiosus* bugs covered in this section can be obtained from The Green Spot by calling 603/942-8925. Detailed release instructions are provided with every order. Here is the current, industry-average pricing for 1998...

Item no. COI5C ... 500 O. insidiosus adults ... 1 unit = $54.60

Item no. COI5C-8 ... 500 O. insidiosus adults ... 8-11 units = $49.14 ea.

Item no. COI5C-12 ... 500 O. insidiosus adults ... 12+ units = $44.23 ea.

Item no. COIN5C ... 500 O. insidiosus nymphs ... 1 unit = $58.50

Item no. COIN5C-8 ... 500 O. insidiosus nymphs ... 8-11 units = $52.65 ea.

Item no. COIN5C-12 ... 500 O. insidiosus nymphs ... 12+ units = $47.38 ea.

whiteflies

Order: *Hemiptera*   Sub Order: *Homoptera*   Family: *Aleyrodidae*

People will often tell us about their pests. They'll sometimes give us obscure descriptions, bizarre morphology details, etc. Often this is done because they aren't quite certain what pest they actually have — and when they're through, very often neither do we. But sometimes we hear a description like this: "and they're like little white flies..." or "like a cloud when the plants are disturbed..." These bits and pieces usually make this pest's identification a breeze — at least generally speaking.

When speaking of whiteflies, at least for the purposes of biological pest control, a general identification is usually not enough — especially if there are budgetary concerns. Proper identification down to genus is really fundamental if you want control at the lowest possible cost. For an explantation read on.

There are three whitefly species of particular concern to growers (although there are more than three species which can be considered problematic): the greenhouse whitefly (Trialeurodes vaporariorum), the silverleaf whitefly (Bemisia argentifolii), and the sweet potato whitefly (B. tabaci).

The first species is probably still the most common —though the others are gaining ground quickly. It is also the easiest and least expensive to treat for. The other two species must be dealt with another way; one which is a tad pricier. The first species can also be dealt with the same way as the latter two, but a cheaper alternative is available for its control. So if you want whitefly control, that's no problem, but if you're on a budget, learn to tell them apart — 80% of the time you'll probably be able to use the less expensive products.

The first species, greenhouse whitefly, can be distinguished from the other two with a simple hand lens. And, by the way, just because the whiteflies are

---

Fig. 8

Help! Which whitefly species do I have?
One of the best ways to tell is to look at the pupae.

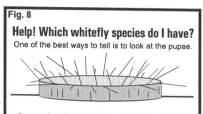

As seen from the side (with magnification only), looking along the leaf's surface, the greenhouse whitefly's pupa (above) looks as if it were made with a tiny cookie cutter; with its sides at 90° to the leaf's surface. It is also fringed with many long hairs.

The sweet potato whitefly's pupa (below), on the other hand,

is more dome-shaped, tapering down to the leaf's surface gradually. The hairs on its fringe are shorter and less abundant. The silverleaf whitefly's pupa looks much the same.

Computer illustration, above and facing page, by M. Cherim

in a greenhouse, has no bearing on what species they are. The easiest thing to do is to locate the pupae or scales, which should be present on the older leaves if the infestation has been around awhile. The pupae of the greenhouse whitefly are raised with sides perpendicular to the leaf's surface and fringed with wild and crazy "hairs." The silverleaf and sweet potato whitefly species' pupae are dome shaped and lack most of the hairs. (See photo, inset above, and figure 8 facing page, for clearer details.)

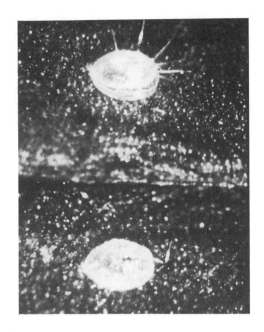

Note the wild Albert Einstein-like "hairs" of the greenhouse whitefly's pupa (top), compared to the more Kojack-ian pate of the sweet potato whitefly's pupa (bottom). The silverleaf whitefly is pretty much the same, with exception to a waxy fringe noticeable under high magnification.

Another way to tell them apart is to look carefully at the adults themselves - though this is the least reliable method. The wings of the greenhouse whitefly appear wider and lay more flat, or are more parallel to leaf's surface. The Bemisia spp., on the other hand, appear thinner, are more tented, and are more at right angles to the leaf's surface. This not a foolproof method of identification, so identifier beware. (See figure 9 right, and the photos on the next page, for clearer details.)

If you had whiteflies last year, but failed to identify them, you should probably consider it this year when, more-than-likely, you'll get them again. Next year, if you have greenhouse whiteflies, you can run a less expensive program — or at least start it that way. It is important,

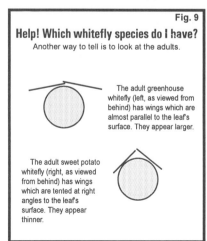

Fig. 9

Help! Which whitefly species do I have?

Another way to tell is to look at the adults.

The adult greenhouse whitefly (left, as viewed from behind) has wings which are almost parallel to the leaf's surface. They appear larger.

The adult sweet potato whitefly (right, as viewed from behind) has wings which are tented at right angles to the leaf's surface. They appear thinner.

Photo by B. Costello, courtesy of Applied Bio-nomics, Ltd.

The greenhouse whitefly is shown in the photo above. The sweet potato whitefly is pictured below. The differences are slight, at best.

however, that you realize that after an infestation is full-blown, it will cost a considerable sum to get it back under control; biologically or chemically. The money savings we keep bringing up is realized during the preventive and/or low infestation phases of the greenhouse whitefly establishment, when the GHWF parasitoid, Encarsia formosa, can be used (see next section for details).

The greenest of scouts won't miss a huge flare-up of whiteflies. Catching them early, however, can take some practice. Doing just the right things is important. First of all, yellow sticky traps should be used to capture early arrivals. If the scout looks at the traps just after some whiteflies have been caught, they will appear, as their name suggests, like white flies (thought they're not actually flies). If the scout doesn't get around to looking at the traps in a timely manner, the white might be washed away by the adhesive to show the insects' true color: orange. The scout may also choose to "beat the bushes," if you will. This will make established populations take flight.

Poinsettias, hibiscus, certain herbs, tomatoes and cucumbers are just a small sampling of what these pests will fall head-over-heels in love with. And they say only good guys wear white hats?

For more detailed information and recommendations concerning whiteflies and ways to effectively control them, you are invited to call The Green Spot, Ltd. at 603/942-8925. ▓

Photo, top, by N. Tonks, photo, bottom, by M. Herbut, both courtesy of Applied Bio-nomics, Ltd.

whitefly controls

The Whitefly Destroyer

Delphastus pusillus
(DEL-fass-tuss POOH-sill-uss)

Description— Like *Hippodamia convergens* (see aphid controls), *Delphastus pusillus*, are ladybug beetles. Exceptionally small ones, but ladybugs nonetheless.

These beetles, with their small, shiny round black bodies (see illustration, inset, and micrograph, below), **love whiteflies**. They can clean up large populations. And, as most beetles are, *D. pusillus* are very opportunistic and will eat pests other than whiteflies: **spider mites, newly born aphids, insect eggs**, etc. We do NOT recommend using these beetles for other pests, though. We

The adult D. pusillus beetle (shown here at 12.5x) is a mere 2 mm. in size. Its appetite, though, is surprisingly large by comparison.

believe control of other pests might not be obtainable with typical predator releases. Coincidental cross-predation should be regarded as a bonus of the application and nothing more.

Built like a tiny tank, these beetles are designed to breech enemiy lines and have been instructed by General Nature to take no prisoners.

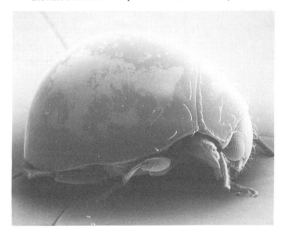

With whiteflies, though, these beetles have proven themselves worthy. Most of our contacts have reported good results with their use. Especially our greenhouse and interiorscape customers.

D. pusillus are shipped as pre-fed, pre-mated, adults (see photo, page 129). These beetles are

Illustration, top, by M. Cherim
S.E.M. photo, bottom, by N. Cherim

shipped with a tiny amount of Biodiet™ in their jar, but it is provided merely as a moisture source. These predators eat very little of the stuff, so it is not recommended for supplemental usage at your release site.

Some popular prey of these beetles includes: the **greenhouse whitefly** (*Trialeurodes vaporariorum*); the **sweet potato or tobacco whitefly** (*Bemisia tabaci*); the **silverleaf whitefly** (*B. argentifolii*); other *Trialeurodes* species; the **Japanese bayberry whitefly** [(*Parabemisia myricae*) as we've had the opportunity to see]; and more obscure, lesser-known species, including, supposedly, the **California giant whitefly (sp. unk.)** which we understand is a new threat in the west.

Life-style— The tiny 2 mm. adult female beetles lay their eggs in the middle of the "rings" of whitefly eggs — up to 75 of them. [Please note: the immature stages of the whitefly, their "scales," which are found sometimes in great numbers on the undersurfaces of the middle to lower leaves often mistaken for the eggs, which are usually located on the undersurfaces of the upper leaves. The eggs are very, very small (0.2 mm.) are quite difficult to actually see.] The eggs hatch into small pale-yellow larvae (see illustration, inset).

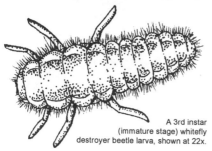

A 3rd instar (immature stage) whitefly destroyer beetle larva, shown at 22x.

These, too, are fierce predators, growing up to 3 mm. long and consuming vast numbers of small, immature stages and whitefly eggs (the larvae, *and* adults, always eat the youngest, most tender morsels first).

The life-span of these predators is roughly 3 weeks in their immature stages, then 4-5 weeks as adults. The conditions for optimum performance will be between 60-90°F with a relative humidity of 75% and up. But these are *optimum* conditions, and not necessarily a prerequisite of successful implementation. Please note, however, cooler temperatures will hamper reproduction and development a certain degree.

Benefits— These beetles have been known to completely devastate a well-established whitefly infestations in a small amount of time. The females demonstrate a keen sense of survival awareness for their young; laying the eggs in the center of whitefly egg-rings assures the young will find food regardless of the direction travelled after hatch, and thus find it easier to survive.

The long-term establishment of these predators are possible in outdoor

Illustration by M. Cherim

conditions, mostly (see Drawbacks). It's tough indoors because it is nearly impossible to maintain an adequate food source. The beetles consume the whiteflies necessary for their establishment too quickly (this is probably going to really upset growers, huh?).

These predators will consume some pollen and a small amount of honeydew (whitefly poop, yuk!). This will aid their proliferation by a small degree.

Searching is the trait which most supports the widespread use of these beetles. They're excellent at it. This allows the release of fewer beetles without compromising results. Especially in large outdoor applications where the hot-spots these beetles are famous for cleaning up are more spread out.

Snug and secure, these beetles are destined for a release in a whitefly-laden environment.

Drawbacks— *D. pusillus*, a Florida native, cannot handle harsh winters, even though light frosts are okay. Consequentially, their ability to establish will be negated in northern climes.

They can't handle too much honeydew. Check out the honeydew situation first. It may be necessary to attempt cleaning heavy deposits.

Some crops, tomatoes for example, have hairs which tend to impede the progress of the *D. pusillus* larvae. This must be taken into consideration when figuring your release rates (see Rates) to compensate for this drawback.

Scouting— Look for whitefly "scales" which look like they've been partially eaten. Finding the adults or larvae can be fairly exhaustive work, though adults may be seen sometimes on overcast days flying around.

Clean new plant growth and natural honeydew reduction may also be apparent scouting indicators.

Monitoring whitefly adult numbers with yellow sticky traps (see the scouting section) may also be a fair indicator. If adult numbers are really high, one strategy we know of (see Fig. 10, next page), can reduce those numbers

Photo by D. Simser

extremely fast, before predator-one is ever released.

Advisories— Aside from misting the site with water before releasing and doing so in the evening (sometimes not necessary in interiors), there are other things you can do to ensure the maximum number of beetles concentrate on the pests at hand. Flowering, pollen producing plants can be a big plus.

Suck 'em up!   Fig. 10

As you might have noticed when your plants are disturbed, whitefly adults take flight. This looks awful, but it really is a great opportunity. Take a vacuum cleaner and have it poised to suck the adult whiteflies right out of the air. Now give the plant a slight shake and go to work. 🔳

Ants, if present, should be controlled. They will defend whiteflies from predators and parasites to protect their honeydew/excrement food, barf! Use barrier products or boric acid products to control the ants.

These beetles may be attracted to light colors, so watch your sticky traps, if you're using them. If you're catching too many beetles, remove the traps from the site or set them out for only 2-3 days per week.

Usages— Greenhouses, interiorscapes, southern orchards, fields and nurseries. Anywhere conditions are right and food is plentiful!

Rates—

```
PRVNT: N/A
CTRL-L: 1-2 / YD. - TRI-WKLY - 2-3X
CTRL-M: 2-3 / YD. - BI-WKLY - 2-4X
CTRL-H: 3-4 / YD. - WKLY - 3-5X
MAINT: 1-2 / YD. - MTHLY - INDEF
GRDN: 40-55% of rates above.
ACRE+: 10-20% of rates above.
COMMENTS: For releases in tomatoes, use
150-200% of rates determined above.
```

Pricing information— The *D. pusillus* beetles covered in this section can be obtained from The Green Spot by calling 603/942-8925. Detailed release instructions are provided with every order. Here is the current, industry-average pricing for 1998...

Item no. CDP1C ... 100 *D. pusillus adults* ... 1 unit = $21.60

Item no. CDP1C-8 ... 100 *D. pusillus adults* ... 8+ units = $18.36 ea.

Item no. CDP250 ... 250 *D. pusillus adults* ... 1 unit = $48.00

Item no. CDP250-6 ... 250 *D. pusillus adults* ... 6+ units = $40.80 ea.

Item no. CDP5C ... 500 *D. pusillus adults* ... 1 unit = $87.00

Item no. CDP5C-4 ... 500 *D. pusillus adults* ... 4-7 units = $80.47 ea.

Item no. CDP5C-8 ... 500 *D. pusillus adults* ... 8+ units = $74.44 ea.

The Greenhouse
Whitefly Parasitoid

Encarsia formosa
(enn-KAR-see-ah FOR-mose-ah)

Description— These 0.7 mm. mini-wasps are best used for preventing the establishment of the **greenhouse whitefly (*Trialeurodes vaporariorum*)**. They can also tackle minor to medium infestations. And, if established, they can adequately protect a crop throughout the season. In addition to the greenhouse whitefly, *E. formosa* can parasitize a few other whitefly species, the sweet potato whitefly (*Bemisia tabaci*), for example. Acceptable prevention and control of other whitefly species, however, is effectively unobtainable with *E. formosa* and we do NOT recommend their use for this purpose.

The parasitized whitefly pupae affixed to these convenient hanging cards resemble ground pepper. These cards should be hung from branches in places where they won't get too soaked, damaged, etc.

Encarsia formosa are supplied as pupae protected in the greenhouse whitefly scales (pupae) which they had used as a host, and actually still are. These parasitized scales are adhered to a small card which can be hung in the plants (see photo, above). The cards come in perforated strips of ten; equal to 1000 *E. formosa*. And when we say 1000 parasitoids, we mean 1000 hatching wasps, not just 1000 pupae (not all will hatch).

E. formosa are also available as loose parasitized scales supplied in a small shaker. For this product we recommend some form of distribution boxes or small cups which can be secured in the crop.

Life-style— *Encarsia formosa*, as parasitoids, work by laying eggs in the 2nd through 4th immature whitefly stages. The wasps' larvae which hatch from the eggs, slowly weaken and kill the developing whiteflies from within

Photo by D. Simser

The parasitoid in this photo is typically difficult to see. Looking for E. formosa adults on the cards just after hatch, due to the contrasting colors, is the likeliest of places to make visual contact. Notice the size of the wasp in relation to the post-hatch pupae in the background.

(*endo*parasitism) causing noticeable changes upon pupation (see Scouting). And each female wasp (they're all females by the way) can do this to up to 200 immature whiteflies!

The life-span of these parasitoids is roughly 3 weeks in their immature stages, then up to a month as adults (see photo, inset). The conditions for optimum performance will be between 68-77°F with a relative humidity of 70% or less. But these are *optimum* conditions, and not necessarily a prerequisite of successful implementation. Please note, however, cooler temperatures will hamper reproduction and development a certain degree. For example: these wasps won't fly when temperatures are below 62°F (see Drawbacks).

Benefits— They're very effective preventive agents capable of small-scale control as well. Consequentially, between their low price and resulting prevention, a lot of money can be readily saved, plus a lot of headaches and plant damage avoided.

Moreover, they are really easy to scout (see Scouting). And they're a great part of an IPM program, with quite a few pesticide tolerances (see Fig. 1 - Biorational Substances Chart, page 6).

There is a lot of information out there about these parasitoids. Combined, growers probably have a thousand years or more of experience with these mini-wasps. There are certainly a lot of success stories from our customers (see Drawbacks).

Drawbacks— Then again, some of our customers have a poor or mediocre results, which, of course, is true of all bio-control agents. If you follow the numbers (see Rates), though, and what pests you're actually dealing with,

Photo by B. Costello, courtesy of Applied Bio-nomics, Ltd.

your story should be one of success.

They can't fly below 62°F, but we're not sure what kind of drawback this really is. As long as daytime temperatures are high enough, you'll be okay. But then again, there are probably be times when the temperatures are down for one reason or another. Perhaps this should be listed in Advisories instead of under Drawbacks. These wasps are also temperamental about light levels (see Advisories).

Honeydew levels, if high enough, can interfere with their performance. A couple of different things can play a role (see Advisories)

Scouting— The scales or pupae of the greenhouse whitefly turn jet black when parasitized (see photo, inset). This is a great indicator. The scales of the *Bemisia tabaci* turn tan-brown when, on occasion, they, too, become a host to *E. formosa*.

Yellow sticky traps might not be able to be used at times (see Advisories).

Advisories— Yellow sticky traps should be removed prior to releasing these mini-wasps. To monitor for thrips, use blue or hot pink traps. If yellow traps must be used for fungus gnats, etc., hang them for only two days per week. To reduce adult whitefly levels, fast, instead of placing about a million sticky traps up, Suck 'em Up! (see Fig. 10, page 130).

Ants, if present, should be controlled. They will defend whiteflies from predators and parasites to protect their honeydew food (the excrement of whiteflies , yuk!). Use barrier products or boric acid products to control the ants.

If your planting doesn't have any ants, check to be sure that the honey-

This whitefly pupae contains a developing E. formosa. This is determined by the color: black, instead of the normal off-white. Notice that the waxy fringe remains, but the wild and crazy hairs are no longer present.

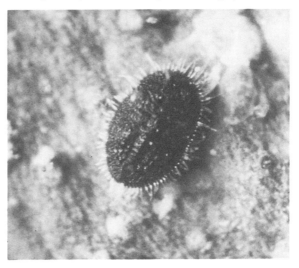

Photo by N. Tonks, courtesy of Applied Bio-nomics, Ltd.

dew isn't too heavy. This may prove to be a hindrance to the parasitoids' performance; they may spend too much time cleaning themselves. Washing the plants with soap and water — at the highest possible pressure — may help reduce the amount of honeydew.

In order for E. formosa to be effective, the release site must be brightly lit, with a minimum of 650 footcandles. Photoperiod or day-length doesn't seem to be of importance, just intensity. By the way, the required light-levels are achieved in normal greenhouse conditions [on a sunny day].

Some crops, due to their excessive whitefly susceptibility, may require special IPM attention, higher release rates and/or more frequently scheduled releases (closer intervals). Greenhouse tomato production is one example, at least according to a couple of our contacts who've suggested we increase the rates for their crop — they felt it was necessary. Another example of a highly susceptible plant would be the eggplant, but it might be used to your favor (see Fig. 11, below for an explanation).

Fig. 11 The Eggplant: A Living Insectary

(A previously published piece by Mike Cherim, edited and revised for this manual.)

Many bio-control users would probably like to rear their own biological pest control agents instead of buying them in. Unfortunately, this would probably be a very expensive practice, unless you had large enough acreage to warrant production.

There is a way to do it for whitefly bio-controls, though. And to twist a phrase: "When in America, do as the Europeans do." Which, as many European growers do, is to put a couple of eggplants in the range [peppers and cucumbers in these case].

The eggplant is a magnet for whiteflies. And having a lot of whiteflies is the key ingredient to whitefly bio-control agent production. Many European growers take advantage of this. They wait until the first greenhouse plants — their eggplants — become slightly infested (all while making small greenhouse-wide Encarsia formosa releases). They then hang a great number of E. formosa cards on the plant — like ornaments on a Christmas tree. This creates kind of a mini-insectary by which the E. formosa can greatly increase their numbers (this is like a trap and banker crop), thus providing protection and control for the rest of the crops in the range.

It is still advisable to release E. formosa in other parts of the range, but the parasitoid numbers, plus the frequency and intensity of the releases can possibly be reduced.

This will also work to some extent with Delphastus pusillus (but to a lesser degree, it should not be introduced directly to the eggplant — it can be too effective), and, perhaps, also Eretmocerus eremicus =californicus nr.

Although this tactic may sound a little scary to most growers, it has been proven in Europe, and now also in Canada. It should work well in several U.S. crops, as well: tomatoes, poinsettias, some ornamentals, in addition to peppers and cukes.

In recent years, advances in bio-control techniques such as this, have come to bear. And surely there will be more to come in our future. We certainly hope so, anyway.

Usages— Most greenhouse crops and brightly lit interior situations. We've seen the successful preventive and light curative implementation of these species in nearly every qualifying situation. We do not recommend the outdoor use of *E. formosa*, with possible exception to small-scale use.

Rates—

```
PRVNT: 4-6 / YD. - BI-WKLY - NEED
CTRL-L: 7-9 / YD. - WKLY - 3-4X
CTRL-M: 10-14 / YD. - WKLY - 3-5X
CTRL-H: N/A
MAINT: 5-7 / YD. - MTHLY - INDEF
GRDN: 45-60% of rates above.
ACRE+: N/A
COMMENTS: For whitefly sensitive crops try
125-150% of rates above plus, PRVNT: WKLY.
```

Pricing information— The *E. formosa* wasps covered in this section can be obtained from The Green Spot by calling 603/942-8925. Detailed release instructions are provided with every order. Here is the current, industry-average pricing for 1998...

Item no. CEFC1M ... 1000 E. formosa on cards ... 1 unit = $10.08

Item no. CEFC5M ... 5000 E. formosa on cards ... 1 unit = $42.00

Item no. CEFC10M ... 10k E. formosa on cards ... 1 unit = $73.50

Item no. CEFC50M ... 50k E. formosa on cards ... 1 unit = $273.00

Item no. CEFC50M-2 ... 50k E. formosa on cards ... 2-3 units = $245.70 ea.

Item no. CEFC50M-4 ... 50k E. formosa on cards ... 4+ units = $221.13 ea.

Item no. CEFS25M ... 25k E. formosa / loose ... 1 unit = $156.00

Item no. CEFS25M-4 ... 25k E. formosa / loose ... 4-7 units = $140.40 ea.

Item no. CEFS25M-8 ... 25k E. formosa / loose ... 8+ units = $126.36 ea.

The Mixed Species Whitefly Parasitoid

Eretmocerus eremicus = *californicus nr.*

(ehr-et-MOSS-ser-uss eh-REM-ee-kuss =kal-ih-FORN-ih-kuss)

Description— These 0.8 mm. mini-wasps are best used for preventing the establishment of the **greenhouse whitefly (*Trialeurodes vaporariorum*) and the *Bemisia* species, *tabaci* and *argentifolii*.** They can also tackle minor to medium infestations. And, if established, they can adequately protect a crop throughout the season.

E. eremicus are supplied as pupae protected in their host whitefly scales (pupae). These parasitized scales are supplied loose in a bottle with bran flakes (see photo, right) which can be distributed throughout the crop.

For this product we recommend the use of distribution boxes or small cups which can be hung in the crop and reused several times.

Life-style— *Eretmocerus eremicus*, as parasitoids, work by laying eggs. But unlike *E. formosa*, which lay their eggs *in* the 2nd through 4th immature whitefly stages, *E. eremicus* females lay their eggs *underneath* those same stages, with a preference for the 2nd instar (stage). The wasps' larvae which hatch from the eggs begin to enter the host and thus slowly weaken and kill the developing whiteflies from the outside-in (*ecto*parasitism). This causes noticeable host changes (see Scouting). And each female wasp (which equals 50-65% of the total release) can do this to up to 150 immature whiteflies!

The 3,000 count unit of E. eremicus =californicus nr.

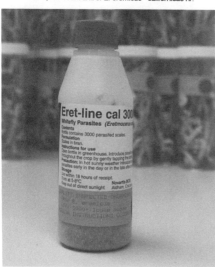

Photo by M. Cherim

The life-span of these parasitoids is roughly 18 days in their immature stages, then for an currently undetermined period of time as yellow-colored adults. The conditions for optimum performance will be between 70-95°F with a relative humidity of 60% or less. But these are *optimum* conditions, and not necessarily a prerequisite of successful implementation. Please note, however, *considerably* cooler temperatures will hamper reproduction and development a certain degree. Higher temperatures, however, don't seem to critically impact the performance of these mini-wasps (see Benefits).

Benefits— These parasitoids, being native to America's desert region, can handle high daytime temperatures, supposedly up to 113°F, as well as cold nighttime temps.

These wasps are easily noticed when developing — though not as much so as *E. formosa* — making scouting a little easier (see Scouting).

Host-feeding activity runs high with this species. It is more than an added bonus; these parasitoids could almost be regarded as predators.

Drawbacks— *E. eremicus* are a bit on the pricey side, at least right now. With increased usage and subsequent demand, however, prices may drop.

Honeydew levels, if high enough, *might* interfere with their performance. A couple of different things can play a role (see Advisories).

We have little practical experience with this species. We, therefore, need your assistance: tell us of your experiences so that we can pass on the knowledge.

Scouting— The scales, when parasitized, begin to reveal the developing wasps inside. Like it is with *E. formosa*, the whitefly pupa does darken as the parasitoid within develops, but not as much so as to conceal the larval wasp.

Yellow sticky traps might not be able to be used at times (see Advisories).

Advisories— Yellow sticky traps might have to be removed prior to releasing these mini-wasps, but we are uncertain and ask that you closely observe the traps if they remain in use. If you need to monitor for thrips, try using blue or hot pink traps. If yellow traps must be used for fungus gnats, etc., you might need to hang them for only two days per week. To reduce adult whitefly levels, fast, instead of placing about a million sticky traps up, Suck 'em Up! (see Fig. 10, page 130).

Ants, if present, should probably be controlled. They will defend whiteflies from predators and parasites to protect their honeydew food (the excrement of whiteflies, ick!). Use barrier products or boric acid products to control the ants.

If your planting doesn't have any ants, check to be sure that the honeydew isn't too heavy. This may prove to be a hindrance to the parasitoids' performance; they may spend too much time cleaning themselves. Washing the plants with soap and water — at the highest possible pressure — may help reduce the amount of honeydew.

Some crops, due to their excessive whitefly susceptibility, may require special IPM attention, higher release rates and/or more frequently scheduled releases (closer intervals).

Usages— On most greenhouse crops and interior situations is where we recommend employing these parasitoids. We are not familiar with any other uses. However, we do encourage you to try them anywhere for the sake of experimentation.

Rates—

```
PRVNT: 6-8 / YD. - BI-WKLY - NEED
CTRL-L: 9-13 / YD. - WKLY - 3-4X
CTRL-M: 14-18 / YD. - WKLY - 3-5X
CTRL-H: N/A
MAINT: 7-9 / YD. - TRI-WKLY - INDEF
GRDN: N/A
ACRE+: N/A
COMMENTS: Should you wish to experiment
with outdoor releases, try 50-75% of rates above.
```

Pricing information—

The *E. eremicus* wasps covered in this section can be obtained from The Green Spot by calling 603/942-8925. Detailed release instructions are provided with every order. Here is the current, industry-average pricing for 1998...

Item no. CEE3M ... 3000 E. eremicus / loose ... 1 unit = $52.00

Item no. CEE3M-6 ... 3000 E. eremicus / loose ... 6-11 units = $48.10 ea.

Item no. CEE3M-12 ... 3000 E. eremicus / loose ... 12+ units = $44.49 ea.

other pests

Other pests are comprised of species from several orders and families

You might not know what they are, but you have trillions of them all over your crops. Are they of importance? Probably yes, since trillions of anything should peak the grower's interests. That's right, we're talking about other pests — not the ubiquitous aphid or spider mite. But off-the-wall stuff.

So far in this book we've discussed common pests and their counterpart controls. In this arena, bio-controls can be an awesome solution. However, take a new pest like the lily leaf beetle (Lilioceris lialii), for example, which has been invading areas of Massachusetts and, at least since an employee discovered this pest in her home garden, New Hampshire, too. What are you going to do? We surely don't know yet. In our experiments with this elongate, 1/4"-3/8" glossy bright reddish-orange beetle, the common bio-controls — at least what was tested while the subjects were still around — did nothing. We even asked some lacewing larvae (see aphid controls) if they would consume some of these beetles' freshly laid eggs in lieu of nothing. Unfortunately, their response was negative.

This is where broad-spectrum chemical and botanical pesticides might come in hand. But only after other, safer options have been explored. Remember, it used to be like that for aphids and such, too, before they were so common.

If you have a few bugs in your greenhouse that you can't identify — especially if it's an open system — don't be too concerned. Unlike the lily leaf beetle, most strange bugs should be of no concern to growers. Unless they are obviously eating your plant material or buzzing you, your help, or your customers, don't worry too much, just observe intently and frequently. If strange critters are eating your plants or laying eggs, get them identified at once. If they're laying eggs directly on your plants, it's advisable to remove the eggs before they hatch, your plant is probably a host for the probable pest. If a bug is present in the jillions, ditto, identify them now — even if they appear to be of no consequence to operations or materials — know what they are; a jillion bugs may do a jillion dollars worth of damage.

Your state university's cooperative extension office is the first place to go for help if your bug books have failed you (which is likely if you're dealing with a strange bugger). Don't be afraid to use the services available to you. Cooperative extension offices and/or your state entomology department will probably be tickled to help you identify a "new" pest. You may be helping them keep track of exotic pests in your state without even being aware of it — this will help other growers too. They may have information about your pest over the phone if you can vividly describe this strange new bug's characteristics. Maybe it's "new" only to you. Who knows, they'll probably have some suggestions on ways to control the bug if it turns out to be a pest after all. If you do know what the new bug on the block is, especially if it's a pest, and even if you do know how to control it, but it's occurrence is not typical, call your state or university contacts, not to get their help, but to help them understand what's going on in your county, state or region so that other growers may benefit from your experience.

The robber fly is one of many insects a grower might encounter. Is it friend or foe?... you might want or need to know.

The robber fly (see photo, below), a member of the Asilidae family, which is 850 species strong just in New England, is typical of an insect which may be seen on its own. Is it a pest? Not knowing, one might kill it just in case. It does look a bit "pesty." But it should have been confirmed. The robber fly is a predator as a larva and as an adult. The larvae, which dwell in the soil or organic material, eat soil pests; the adults will attack moths, flies and numerous other critters in mid-flight. Heck, the robber fly's so bad, it's good!

The next section will detail some bugs we feel are out of the typical pest control loop. The first species is a pollinator, the next three are parasitoids and the final one's a showy but commercially ineffective predator. They differ from the other beneficial organisms listed in this book because of either their tastes, nature or role. Read the next section to understand what we mean, specifically. Excuse the format; it, too, is different.

For more detailed information and recommendations concerning other pests and ways to effectively deal with their presence, you are invited to call The Green Spot, Ltd. at 603/942-8925. ✄

Photo by M. Cherim

other bugs

Bumblebees
as Pollinators

Bombus impatiens
(bom-BUSS IM-pay-shinz)

General information— Greenhouse tomato plants (as just one example), as do their outdoor brethren, require pollination to fruit. Outdoors this is done by bees, wasps, wind and other natural sources. In the greenhouse, however, people and sometimes their people-made devices must be the bees and the wind. On tomato plants, growers manually, or with a small "buzz" device, shake or vibrate the flower laden trusses (the flowers have sex and tomatoes are born).

This is all fine and dandy, it works quite well — many tomatoes are born each day. The problem is the amount of focused manpower needed to properly carry out the pollination process. That's where commercially produced bees come in: commercial bumblebee hives, for example, are designed exclusively for crop pollination. Not just for tomatoes, but other crops as well: peppers; cukes; squash; cane-, straw-, blue- and cranber-ries; and many other crops in need of primary or supplemental pollinators.

Growers should use bumblebees shortly after the first flowers appear.

This bumblebee is visiting the flower cluster of a catnip plant. Note the large collection of pollen attached to the bee's hind leg.

Photo by M. Cherim

Normal greenhouse conditions [without the normal greenhouse sprays (see Fig. 1 - Biorational Substances Chart, page 6)] can support the hive, though excessively high temperatures may hamper activity. Cool temperatures, on the other hand, may slow activity, but don't completely halt it. That's the nice thing about bumblebees versus honeybees, they can fly in cool weather, down to 41°F, even in the rain and on windy or cloudy days, visiting flowers, bruising them (a scouting sign of activity), doing their thing. Honeybees won't even come out of their hive in those conditions, and temperatures have to be at least 50°F. Consequently, *B. impatiens* are a viable option for outdoor use in addition to greenhouse applications. They're bigger, faster, better and more versatile.

Please bear in mind, though, not the *bees*, but the actual *hives, must* be sheltered from the elements if used outdoors. A half-sheet of plywood will do. And indoors, the hives must be placed above the ground, in a vibration-free (from fans, etc.) place where they will be sheltered from irrigation water, traffic, etc., yet open from all frontal angles of approach. Placed on 2 cinder blocks on end, perhaps, and in such a way that ants cannot access it (by applying a sticky trap compound to the blocks). By the way, don't open the little doors until the bees inside settle down. They can be agitated from all the handling and might respond to your handling aggressively, so wait at least one hour.

In fact, anytime while using bees, it should be noted that **they CAN sting. If you're hypersensitive to stings, do not go anywhere near these hives.** It is, in fact, best to not even enter a greenhouse or field while the bees are being used. The non-allergic individual should go about his or her business as usual, but occasional stings can occur. Give the bee hive a wide berth when possible. Additionally, try not to make wild movements around the hive, as this may evoke aggression. Bright colored clothing may spark curiosity from the bees, but should not provoke them as long as you keep your composure. In a nutshell, bumblebees should be given due respect. They are not domesticated or trained, just laboratory reared.

The hive is constructed of cardboard and plastic and comes with its own food source. The hive is completely self-contained. Within the hive comes one of two different colony sizes: a Class 'A' colony, or a Class 'B' colony.

The Class 'A' hive has a queen and a queen larva and will last for approximately 10-13 weeks. This size treats up to 20,000 square feet.

The Class 'B' hive has a queen and will last for approximately 8-10 weeks. This size treats up to 8,000 square feet.

If you, like many growers, have several small greenhouses, instead of one 8,000 square footer, for example, you may encounter difficulty keeping them pollinated. That's were the "hotel hive" comes in. It is placed in a greenhouse while some of the present hive's bees are out foraging. The active hive is replaced by the hotel hive. It is then moved to another greenhouse while the hotel hive awaits the remaining foragers, it's their new and temporary home.

Bombus impatiens can be used in all states except AZ, CA, CO, ID, MT, NM, NV, OR, UT, WA and WY. For the listed states use the other commercially available species: *B. occidentalis*. Call for more information about the western species.

Most growers have success with the Natupol hives. Most growers don't like the fact that they have to pick up the hives at the nearest airport (they are currently shipped via commercial airline). However, with the huge labor-savings and improved fruit-set, nobody's complaining too loudly.

Purchasers of hives please note: These hives are shipped directly from the producer. We, therefore, cannot regulate the product's guarantee. If you experience problems with a hive, call *us* right away, and we'll try to get a replacement for you if that's what's needed to fix the problem. Understand, however, that you will often be billed for the replacement's freight if you wait too long to notify us of the problem. This is the *producer's* policy.

Pricing information— The *B. impatiens* bumblebees covered in this section can be obtained from The Green Spot by calling 603/942-8925. Detailed release instructions are provided with every order. Here is the current, industry-average pricing for 1998...

Item no. DBBHB ... Class 'B' B. impatiens hive ... 1 unit = $145.00

Item no. DBBHB-2 ... Class 'B' B. impatiens hive ... 2-3 units = $140.00 ea.

Item no. DBBHB-4 ... Class 'B' B. impatiens hive ... 4+ units = $135.00 ea.

Item no. DBBHA ... Class 'A' B. impatiens hive ... 1 unit = $245.00

Item no. DBBHA-2 ... Class 'A' B. impatiens hive ... 2-3 units = $240.00 ea.

Item no. DBBHA-4 ... Class 'A' B. impatiens hive ... 4+ units = $235.00 ea.

Item no. DBBHH ... Hotel hive ... 1 unit = $35.00

Item no. DBBNS ... Nectar solution [feed bag] ... 1 unit = $15.00

A Filth-Breeding Fly Parasitoid Mix

Muscidifurax raptorellus, M. zaraptor & Spalangia endius

(mew-sid-ih-FURE-ax RAP-tor-ell-uss, M. zah-RAP-tor & SPAH-lan-gee-ah ENN-dee-uss)

General information— A serious drawback to raising livestock is the flies. All sorts of filth-breeding flies make their home in the manure present with such an activity. Well, we have a solution, in addition to regular manure management and trapping, try using a filth-breeding fly parasitoid mix.

These mini-wasps are shipped as pre-parasitized fly pupae mixed in a bag containing a small amount of pine shavings as a carrier. Upon receipt you can mix the bag's contents with more shavings so that you may find it easier to treat a larger distribution area.

In normal stable conditions, with a preference to the dry side, and only if programmed releases are followed, many common species of flies can be adequately controlled with these nocturnal (active at night) mini-wasps (assuming other tactics are practiced, as mentioned above).

Here a fly parasitoid is ovipositing [laying eggs] in a fly pupa.

The wasps work by laying up to 5 eggs in each fly pupa (see photo, above). The wasps' eggs hatch and the larval offspring consume the host [developing fly] from within (endoparasitism). New parasitoids emerge from the host, instead of a new fly. Thus the cycle continues.

Photo by M. Badgley, courtesy of Buena Biosystems, Inc.

Sprinkle the bag's contents in and around stalls, feeding areas, manure storage sites and, in other, non-livestock worlds, compost piles, refuse disposal sites [landfills], etc. Put them in protected locations (from hooves, rain, etc.).

Start very early in the year (early- to mid-spring) with releases every 3-4 weeks, then close the interval between them as the fly season goes into full-swing. It usually peaks in July-August. If you don't start early you'll have a tough time trying to catch up — you'll have to make a couple double-size applications, at least. The reason is the flies reproduce faster than the parasitoids.

Many of our contacts have great success with these parasitoids — though some, of course, do not. *Most* are pleased. Again, the mini-wasps are just part of the program; trapping and sound manure management are essential. The successful ones are those who ask us to put them on a regular shipment schedule. This works the best. There's no forgetting, and we make sure the timing is right.

Farms, dairies, kennels, dumps, race tracks, etc. Use these wasps anyplace flies are a problem. Try the following formulas, they should make it easier to determine what you'll need per release:

Multiply your total livestock weight by 1.6 (= no. needed)
or, for other applications, use 65-85 wasps per cubic yard of compost, etc.

Pricing information— The fly parasites covered in this section can be obtained from The Green Spot by calling 603/942-8925. Detailed release instructions are provided with every order. Here is the current, industry-average pricing for 1998...

Item no. CFPM5M ... 5000 Fly parasite mix ... 1 unit = $18.00

Item no. CFPM10M ... 10k Fly parasite mix ... 1 unit = $25.20

Item no. CFPM25M ... 25k Fly parasite mix ... 1 unit = $49.87

Item no. CFPM50M ... 50k Fly parasite mix ... 1 unit = $73.50

Item no. CFPM100M ... 100k Fly parasite mix ... 1 unit = $122.85

Item no. CFPM100M-2 ... 100k Fly parasite mix ... 2-3 units = $110.56 ea.

Item no. CFPM100M-4 ... 100k Fly parasite mix ... 4+ units = $99.51 ea.

The Praying Mantis

Tenodera aridifolia sinensis
(ten-oh-DERR-ah ah-RID-ih-fole-ee-ah sih-NEN-siss)

General information— Praying mantids (see illustration, below). Some say they're valuable "general" predators. Well, they're certainly not picky about what they eat (and they *do* eat ladybugs, contrary to popular thought). But as far as *valuable*, as commercial bio-control agents, anyway, they are NOT. Praying mantises are too territorial and cannibalistic. Hunt and eat, yes; protect your crops from aphids, whiteflies and mites, no way!

Adult praying mantis

"Then why," you may ask, "do bug purveyors sell them?" Some sell them to make the bucks, and will tell you anything you wish to hear to make them seem the ideal choice. For the reputable companies, the answer can be summed up in two words: *They're awesome!*

The mantises are supplied as egg-cases, one egg-case per hatching bag. The yield of an egg-case is normally between 50-400 baby mantids — each about three-quarters of an inch long (see micrograph, below). They're available from Jan. to June but, for best results, try to get them just before the last frost in your area.

Other uses for these insects include their showiness. Like ladybugs, people know what they are. They can be symbolic of bio-control and IPM. Present in a garden

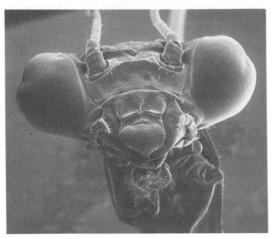
First instar mantid shown at 51x.

center's greenhouse, mantids can help tell your customers about the Green Methods you're using. In a garden center's greenhouse they can also draw the attention of young people, distracting them as their parents select plant materials (sort of a biological kid control agent).

Mantises are territorial stalkers of insects — good and bad. They'll spend days tracking and waiting

Illustration, top, by M. Cherim
S.E.M. photo, bottom, by N. Cherim

for a fly (see photo, right). The fly finally gets close enough and the mantid's serrated forelegs strike out at 1/20th of a second, capturing it. The mantis delivers a bite to the neck, killing the fly, then savors its meal. That's how they work, that's why they're cool. Now shouldn't you order one or two for yourself?

A fourth instar mantid scouting clover.

Pricing information— The praying mantis egg cases covered in this section can be obtained from The Green Spot by calling 603/942-8925. Detailed release instructions are provided with every order. Here is the current, industry-average pricing for 1998...

> **Item no. CTAS1 ... Praying mantis egg-case ... 1 unit = $2.59**
>
> **Item no. CTAS1-12 ... Praying mantis egg-case ... 12-23 units = $2.07 ea.**
>
> **Item no. CTAS1-24 ... Praying mantis egg-case ... 24-95 units = $1.66 ea.**
>
> **Item no. CTAS1-96 ... Praying mantis egg-case ... 96+ units = $1.33 ea.**

Other, Other Bugs

The Green Spot can also provide information about the following organisms. Please call for details, pricing and availability.

Anaphes iole, Aphelinus abdominalis, Bombus occidentalis, Eisenia foetida (call for source), Encarsia formosa-BTS (Beltsville strain), Galendromis occidentalis, Leptomastix dactylopii, Mesoseiulus longipes, Neoseiulus californicus, Pediobius foveolatus, Podisus maculiventris, Phytoseiulus macropilis, Thripobius semileuteus, Trichogramma platneri and Trichgrammatoidia bactrae. PLUS, the eggs of the grain moth, Sitotroga cerealella, which are useful in several bio-control projects, are available either fresh or frozen. Please call 603/942-8925.

Photo by M. Cherim

Other Products Used

a close-up look at the tools at hand

photo, opposite, by M. Cherim

education

Educational resources are the grower's first step

Books, lectures, catalogs, the internet, growers' meetings, trade magazines, discussions with industrymates, even late night bs-ing at a bar: they're all good sources of educational enrichment — some, perhaps, better than others. So what does this have to do with bio-control and integrated pest management (IPM)? Everything. Boning up, if you will, is really fundamental to pest control success.

Now if you're thinking that if you use chemicals and other conventional methods of pest and disease management, you don't need bone up on the latest trends and practices, think again. Technology is rapidly advancing in many areas, perhaps all areas. It's best to stay abreast of many topics.

Written material is positively the most accessible form of enrichment. A book is truly portable. Access is instantaneous. And as covers weaken with age, an intimacy develops. Lines, paragraphs, pages, even chapters, are known by rote. And think of this: what friend would so willingly accompany us to the privacy of the restroom, or tuck us in and lull us to sleep each night, as would a book?

In the particular section which follows, a few favorite titles are detailed. Check them out — the answers to your question are likely within.

For more detailed information and recommendations concerning the betterment of your bio-control and IPM education, you are invitied to call The Green Spot, Ltd. at 603/942-8925. ⌗

educational goods

The Green Methods Manual
The Original Bio-control Primer - Edition IV for 1998
by Mike Cherim - 1998 ⊂⋂⊃∪

This paperback contains information derived from actual growers and others intimate with bio-control. The best way for us to describe it to you is to have you describe it to us. You're holding it right now, so we hope you're enjoying it, or at least finding it useful. Features 134 quality B/W photos. 244pp.

Item no. EGMM4 ... 1 manual ... Each = $9.95
Item no. EGMM4-6 ... 6-11 copies ... Each = $8.21
Item no. EGMM4-12 ... 12-23 copies ... Each = $7.56
Item no. EGMM4-24 ... 24 or more copies ... Each = $6.22

Applied Bio-nomics, Ltd.
Biological Technical Manual - Edition II
by Don Elliott - 1997

Coming soon, we hope. Photo and copy relevant to Edition I. Call for Avail. ⊃

An excellent publication focusing on the specific uses, biology, application, pesticide compatibility, etc., of 10 biological pest control agents.

This 3-ring bound manual includes specific bio-control program information on several greenhouse crops: cucumbers; cut lisianthus and stocks; cut and potted gerberas; gloxinias and African violets; poinsettias; and tomatoes. It also has sections on general floriculture crops and interiorscapes.

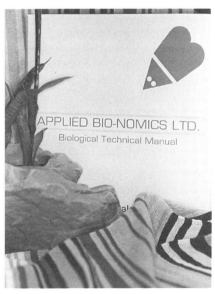

Photo by M. Cherim

Contains specific information about 5 common pests: aphids, fungus gnats, thrips, two-spotted mites and whiteflies. Features a pullout section with 19 full-color photos of common pests and their natural enemies. 90pp.

Item no. EABBTM ... 1 manual ... Each = $38.95

Ball Pest and Disease Manual -
Edition II
↻ by Charles C. Powell, Ph.D. & Richard K. Lindquist, Ph.D. - 1997

This fine hardcover features specific information about plant pathology, entomology, and the interaction between the two, in flower and foliage crops. Features 68 full-color photos, plus 128 quality B/W photos. Has detailed diagnostic and corrective information. A very complete publication. 426pp.

Item no. EBPDM ... 1 manual ... Each = $57.95

Common-Sense Pest Control
by William Olkowski, Ph.D., Sheila Daar & Helga Olkowski - 1991 ↻

"excellent reference, worthy of space in any library"
—The New York *Times*

This extremely detailed, hardcover publication provides information about integrated pest management practices in the greenhouse, on the farm, in the garden, in the community, in the home and even on the body. It is probably the most complete compilation of least-toxic pest control methods available anywhere. Very thorough; a must for everyone. Features 55 B/W photos, 329 quality illustrations and 110 charts and tables. 720pp.

Item no. ECSPC ... 1 book ... Each = $39.95

Both photos, courtesy of respective publishers

Rodale's Controlling Pests and Diseases

☞ by Patricia S. Michalak & Linda A. Gilkeson, Ph.D. - 1994

This paperback, one of the Successful Organic Gardening Series books, is a good starter book. It is full of basic information in a very easy-to-digest format. It provides information about 130 common pests and diseases, plus basic techniques to organically control them. It features an incredibly vivid, full-color photo collection (248 shots) and several detailed illustrations. Our first choice for the coffee table. 160pp.

Item no. ERCPD ... 1 book ... Each = $14.95

Rodale's Pest & Disease Problem Solver

by Linda Gilkeson, Pam Peirce and Miranda Smith - 1996 ☞

This hardcover book may rival or even surpass Rodale's longtime best works (which is still our favorite, we think). Like the title which follows (Organic Gardener's Handbook of Natural Insect and Disease Control), this book reveals a plant-by-plant guide to problem solving and prevention, an insect identification guide and a disease symptom guide. This book is nicely balanced with over 300 full-color photos designed to make it even more informative. It's easy to locate problems *and* solutions. It utilizes multiple channels to help you locate the information you need. This book may be the new one which gets more use than any other book on your shelf, unless the book covered next still gets that honor as it has in the past. Designed for home gardeners, but useful to, and needed by everyone. 384pp.

Item no. EPDPS ... 1 book ... Each = $27.95

Photo, top, by M. Cherim
Photo, bottom, courtesy of publisher

The Organic Gardener's
Handbook of Natural Insect and Disease Control

♨ A Rodale Press Garden Book edited by Barbara W. Ellis & Fern Marshall Bradley - 1992

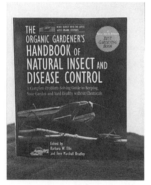

This paperback is probably Rodale's best work (or was). It reveals a plant-by-plant guide to problem solving and prevention; an insect identification guide with 262 full-color and 2 B/W photos; a disease symptom guide with 96 full-color photos; and a section on various controls. It also contains several color and B/W illustrations. Very well laid out, and very easy to find topics in it. Utilizes several avenues to help you locate the information you need. This book will probably (or used to) get more use than any other book on your shelf — if you can (could) ever *keep* it on your shelf, that is. Designed for home gardeners, but useful to, and needed by, every grower employing or contemplating [almost] organic methods. 328pp.

Item no. EOGHNIDC ... 1 handbook ... Each = $17.95

Color Handbook of Garden Insects

by Anna Carr - 1979 ♨

This excellent paperback book contains 346 large full-color photographs, 1 black-and-white photo, 12 tables, 51 line drawings, 74 range maps, and enough information to make anyone feel comfortable with their knowledge of arthropods.

Information includes the basics like what an insect is and how it lives and grows, and how to provide *some* control, to more thorough information such as range, description (incl. size), life-cycle, host plants and feeding habits. Moreover, it provides basic information to the pests' natural enemies and natural control options.

This publication may not be the most accurate text out there, nor will it be the most complete. And it's definitely not the most up-to-date (it *was* pub-

Both photos courtesy of respective publishers

lished in 1979), but it *is* a worthwhile buy. 241pp.

Item no. ECHGI - 1 handbook = $16.95

The Gardener's Guide to Plant Diseases

by Barbara Pleasant - 1995 ➲

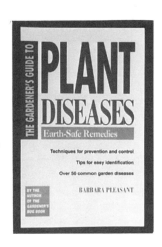

Considering that this paperback book doesn't contain a single photograph (it does have 22 line drawings), it is remarkably informative concerning the causes, identification and earth-safe remedies and prevention techniques of over 50 plant diseases. Additionally, this book is loaded with charts and tables, making its information more accessible. It even has a plant-by-plant table. Well worth the money. 188pp.

Item no. EGGPD ... 1 guide ... Each = $12.95

Knowing and Recognizing

by M. Malais and W.J. Ravensberg - 1992 ➲

This paperback book's subtitle "The biology of glasshouse pests and their natural enemies" defines well its contents. Outstanding information for commercial greenhouse growers — especially European growers (because it's still in Celsius) — enhanced with 77 quality photos.

This book happens to be the official publication of Koppert Biological Systems. It is an excellent book which is professionally laid out and easy on the eyes.

The Koppert people should be proud. This superb publication is a must buy. We strongly recommend it to all of our customers. 109pp.

Item no. EKKR ... 1 book ... Each = $39.95

Photo, top, courtesy of publisher
Photo, bottom, by M. Cherim

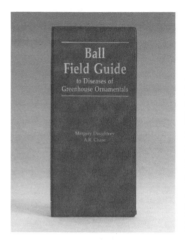

Ball Field Guide to Diseases of Greenhouse Ornamentals

☞ by Margery Daughtrey and A.R. Chase - 1992

It's the soft plastic bound book the pros use. Disease management is closely related to pest control. That's why we try to address disease concerns. And this book does well for us in the information end. It contains 506 full-color photos which accurately detail the diseases associated with 78 ornamental plants! This book is organized by plant, through the table of contents, and by disease, through the index. 218pp.

Item no. EBFDGO ... 1 book ... Each = $67.00

The Color Atlas of Pests of Ornamental Trees, Shrubs and Flowers

by David V. Alford - 1995 ☯

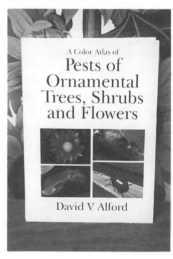

This outstanding hardcover book contains over 1000 large full-color photographs.

This atlas discusses the principals of pest control of ornamental plants, followed by sections on various pests. Each major order and family is considered with details of their status, host range, world distribution, diagnostic features and biology.

This publication describes the characteristic damage caused by a multitude of pests, in addition to some recommended control tactics. A must buy. 448pp.

Item no. ECAPO ... 1 atlas ... Each = $98.95

Photo, top, courtesy of publisher
Photo, bottom, by M. Cherim

scouting

A good scout should be well compensated

There is, perhaps, no individual more responsible to the success of a crop than the lonely scout. No one is certainly in a better position to save the company money. This should be of even greater interest if the company, the grower and the scout are one.

Nearly all major infestations begin with just one or two individual pests. They mate, lay eggs, then, before you know what hit you, your operation is grossly infested. Catching the one or two individuals early, before they have relations, can save you from having to deal with massive numbers, and associated losses, down the road.

Scouting, as a word, speaks little of what it entails. It is, however, the label put on a methodical, intensive process. Without having it spelled out to a degree, though, many growers are unsure of what the process is, specifically, and tend to poke around the operation in a haphazard, irregular fashion whenever there is time. This is not effective scouting, thought it can't hurt to give a look-see now and then, there is a better way.

What follows is our recommendation for setting up and using an effective scouting program that will not only provide early insights, but will offer a significant amount of historical information.

THE GREEN SPOT'S RECOMMENDED SCOUTING PROGRAM

STEP 1: Measure your greenhouse, planting area, field, etc., and draw a scale-size (i.e. 1/8" = 1 foot) map of the area. [For the purposes of this example, a greenhouse will be used.] Indicate on the map the name, number or location of the structure, in addition to its usable size in square feet and other information which is unlikely to change. Include blanks for weekly information such as crop, inspector, date, etc.

STEP 2: In the drawing, detail all permanent structures such as greenhouse benches, etc. (You may want to number the benches.) Include non-production areas, as well. For example, a potting bench should be shown (but probably blocked out or shaded since it will not be part of the usefulness of the map).

STEP 3: Divide and identify the production areas of the greenhouse into 250 sq. ft.

sections. Now you're ready to photocopy the map (see Fig. 12, below) onto white paper. Use one copy each week as a guide to your regimen.

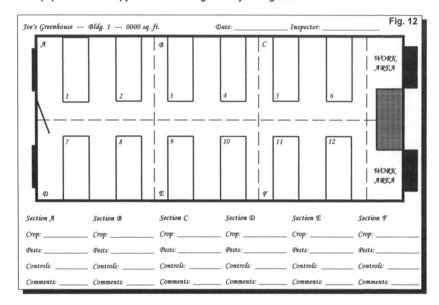

STEP 4: Now that you have a guide for scheduled and organized scouting, it's time to put it to use. First you'll want to make a scheduled time for scouting. Try for one day during the week, say Wednesday, at the same time every week, perhaps from 10:00 am - 11:00 am. This is a good example time, the area's well lit, warm, and most pests should be fairly active. Now you'll want to make sure you get in there to scout on the assigned day, during the assigned time. Fluctuations will cause misreadings and foul your record-keeping.

STEP 5: Following your map, tour the greenhouse in an organized fashion; proceed from the entrance to Section A, check the 1 or 2 sticky trap(s) in that 250 sq. ft. section, check the tagged "indicator plant(s)," which should be representative of the plant material(s) in that section, from top to bottom. Jot down your findings (pests, diseases, controls used, etc.), then move to Section B and continue.

STEP 6: Graph your findings by date, file the map, and get a very real picture of the pest and disease activity in that section for use later. Proper records will give a very clear look at your operation's activities.

VALUABLE REMINDERS

A) Use one trap per 250 sq. ft. section, mounted vertically, 2-3" above the plant

Computer illustration by M. Cherim

canopy.

B) Use one additional trap per 250 sq. ft. section, mounted horizontally very close to the media to detect fungus gnats, etc. (Or use a 1/4" slice of raw potato to capture larvae.)

C) When doing your plant inspections, try shaking the plant over the back of the white scouting map; this will help you greatly in finding pests like thrips.

D) Don't lighten up with the regimen: miss 1 day and you'll taint two weeks worth of records (instead of two 7 day weeks, you'll end up with one 6 day week and of 8 day week).

E) Change your traps if they have such a number of pests that it will make counting difficult or inaccurate the following week. Change them once a month even if they are clean so as to be sure they remain viscous and functional.

Certain materials which are useful to the scout, such as sticky traps, are discussed in more detail in the section which follows.

For more detailed information and recommendations concerning pest and disease scouting techniques and practices, you are invited to call The Green Spot, Ltd. at 603/942-8925. 🞨

scouting goods

Blue and Yellow Sticky Traps

Bugs would do well to comprehend the awesome power of goo. The trap pictured is one of the newer types from biosys. Securing the trap is one of the metal trap stakes.

These traps are one of the scout's most important tools. Aphids, fungus gnats, leafminers, shore flies, thrips, whiteflies and a host of other things (see photo, inset) are all attracted to yellow sticky traps and will land on them (the blue traps are for thrips only). And once they do, they get stuck.

The scout should check the traps at least once a week and count the number of pests trapped. The counts obtained will correlate directly to the severity of the infestation. The traps should be changed at least monthly to assure the adhesive remains viscous (but changed weekly, if they have captured insects, for easier counting).

The primary function of the sticky traps is not to control pests, but to only monitor pest type(s) and levels present in the growing area. However, in *some* cases, reasonable control of fungus gnats and whiteflies *may* be obtained. Even if control is not obtained, yellow sticky traps *will* reduce pest numbers. Unfortunately, they may also reduce the number of good bugs.

If at any point in time more beneficials are getting caught than pests, removal of the traps may be warranted. Normally, however, the advantages of employing these devices far outweighs the disadvantages. Removing the traps does not necessarily mean to do so entirely, but to limit their exposure to the good bugs to one or two days a week

Photo by M. Cherim

Used most commonly are the 3"x5" traps. They're less obtrusive and easier to make your pest counts with. If you desire traps with a larger surface area larger traps are available. Or, if you want, you can make your own. If you want to make your own, see Fig. 13, below.

To use sticky traps, peel back the paper to expose the sticky substance. Then hang the trap in place. A paper clip or clothes pin works well. Yellow sticky traps should normally be placed vertically, 2-3" above the plant canopy. In this position they will capture *most* common pests. When using these traps for fungus gnat monitoring, peel back the paper on only one side of the trap and lay it on the medium, sticky side up. When the useful life of the sticky side has expired, peel the paper from the other side, affix the removed paper to the used side, flip the trap over and put it back into position. For seedlings, low growing crops or small potted plants, use wire trap stakes (see photo, facing page). These reusable, 12" metal stakes are a convenient way to set out the traps.

Please note: Regardless of where you place the traps (using 1 per every 250 square feet), position them in such a way that your plants' leaves, employees or customers will not come into contact with them. If by chance, when placing or checking the traps, you do get the sticky compound [a long-lasting, viscous, polybutene-naphtha inert rubber polymer] on you, a common, waterless hand cleaner will effectively remove it. *Or* you may opt for the newer biosys traps. Their sticky stuff doesn't come off on you. Now that's a product you can stick with (instead of stick to).

When receiving biological pest control agents, check them out, learn to recognize them so if you see them on your sticky traps later, you won't confuse them for a pest. Knowing what the bad critters look like is equally important. In the Educational Goods section we describe some books which may assist you in spotting and recognizing pestiferous species. We've seen firsthand how important proper identification can be: one of our contacts spent hundreds of dollars on mealybug controls, but when results of the program were unsatisfactory, he sent us a sample. The "mealybugs" were actually a wooly aphids — a common, unfortunate mistake. To help the scout what bugs are on the traps, use the guide (Fig. 14) on the next page.

MAKING STICKY TRAPS   Fig. 13

1. From a piece of 1/8" acrylic sheet, or similarly smooth material, cut out your trap shapes.
2. Drill a hanger hole at the top of the traps, if needed.
3. Paint one or both sides with Rustoleum yellow #659.
4. Allow paint to thoroughly dry.
5. Sparingly apply Tanglefoot Trap Coating (offered in this section), or similar material, to the painted surfaces and carefully place the traps.

Fig. 14 - **What's on your traps?**

Aphids— *Normally, only winged aphids will be caught on sticky traps. In most cases, the aphid's wings will settle to either side of its abdomen. Often, the aphid will bear young before dying. The brood, which will be barely visible, will also be trapped. An aphid's front wings have two parallel veins close to the wing's leading edge. The legs and antennae are long and thin. Another distinguishing characteristic is the presence of siphoons (two hornlike appendages at the rear of the aphid).*

Fungus gnats— *More often than not, fungus gnats will be in a multitude of positions because they struggle to free themselves. These small, fragile grey/black insects are mosquito-like, but smaller. They have a distinctive Y-shaped vein close to the outboard edge of each wing (the stem of the "Y" will point towards the body). The legs are very long and thin. The much segmented antennae are also very long and thin, and appear beaded.*

Leafminers— *These pests are small flies and look accordingly. Like fungus gnats, leafminers will thrash about on the sticky trap and become even more mired. They have medium-length legs and short antennae. Another important identifier is a yellow spot on each side of the thorax.*

Shore flies— *These too, as their name implies, look like flies. They are about the size of fruit flies, but are much darker and have black eyes. They have pale spots on their wings, medium-long legs and short antennae.*

Thrips— *These barely discernible insects have long, thin bodies. Their wings, which will usually be closed tightly to the body, are fringed with tiny hairs. The legs will not be visible, and the stubby, thick antennae will protrude from the head to form a "V" shape.*

Whiteflies— *These tiny, fragile flies are normally covered with a white, powdery bloom. When caught in a sticky trap, any of this white coating which contacts the adhesive usually dissolves into the morass, making visible their true color: dull orange. However, body parts which do not contact the adhesive will remain white. Their legs and antennae are medium-short. The whitefly is the easiest pest to identify on yellow sticky traps.*

Other— *If these traps are used outdoors or in an open-system greenhouse — especially one not on a spray regimen — hundreds of different species will be caught on sticky traps. Most of which will be of no concern to anyone. Many will be beneficial to agriculture. If large numbers of a single species appear on your traps, have it checked out. Contact us right away, consult one of our books or notify your cooperative extension entomologist.*

The sticky traps covered in this section can be obtained from The Green Spot by calling 603/942-8925. Here is the current pricing of these products for 1998...

Item no. PMTS ... 12" Metal trap stakes ... Each = $.40
Item no. PMTS-100 ... 12" Metal trap stakes ... 100 box = $34.50

Item no. PST25-Y ... 3"x5" Olsen yellow traps ... 25 pk. = $6.95
Item no. PST25C-Y ... Cs 3"x5" Olsen yellow traps ... 26/25 pks. = $162.18

Item No. PST25-B ... 3"x5" Olsen blue traps ... 25 pk. = $7.30
Item no. PST25C-B ... Cs 3"x5" Olsen blue traps ... 26/25 pks. = $168.67

Item no. PBST30-Y ... 3"x5" Yellow biosys traps ... 30 pk. = $10.83
Item no. PBST30B-Y ... Bx 3"x5" Yellow biosys traps ... 4/30 pks. = $36.10
Item no. PBST30C-Y ... Cs 3"x5" Yel. biosys traps ... 16/30 pks. = $138.62

Item no. PBST30-B ... 3"x5" Blue biosys traps ... 30 pk. = $11.37
Item no. PBST30B-B ... Bx 3"x5" Blue biosys traps ... 4/30 pks. = $37.90
Item no. PBST30C-B ... Cs 3"x5" Blue biosys traps ... 16/30 pks. = $145.55

Item no. PBST10-Y ... 4"x9.5" Yellow biosys traps ... 10 pk. = $5.61
Item no. PBST10B-Y ... Bx 4"x9.5" Yel. biosys traps ... 12/10 pks. = $59.40
Item no. PBST10C-Y ... Cs 4"x9.5" Yel. biosys traps ... 48/10 pks. = $213.84

Item no. PBST10-B ... 4"x9.5" Blue biosys traps ... 10 pk. = $5.89
Item no. PBST10B-B ... Bx 4"x9.5" Blue biosys traps ... 12/10 pks. = $62.37
Item no. PBST10C-B ... Cs 4"x9.5" Blue biosys traps ... 48/10 pks. = $224.53

Item no. PBSTRL-Y ... 4"x100' Yellow biosys trap roll ... 1 roll = $37.50
Item no. PBSTRLC-Y ... Cs 4"x100' Yel. biosys trap roll ... 10 rolls = $295.00

Item no. PBSTRL-B ... 4"x100' Blue biosys trap roll ... 1 roll = $39.37
Item no. PBSTRLC-B ... Cs 4"x100' Blue biosys trap roll ... 10 rolls = $309.75

Tanglefoot Insect Trap Coating

Coat any nonporous surface with this compound and you'll have a trap. This product is great for making and re-coating your own sticky traps. Tangle-Trap® Insect Trap Coating is colorless, odorless and very long lasting — even when exposed to the elements.

Brush Tangle-Trap Insect Trap Coating onto greenhouse bench legs and other surfaces to control ants. To control larger crawling pests such as

Artwork courtesy of manufacturer

caterpillars, use Tree Tanglefoot Pest Barrier, which will also provide excellent ant control (discussed later in this manual). The products discussed here can be obtained from The Green Spot by calling 603/942-8925. Here is the current pricing of these products for 1998...

Item no. PITC8 ... 8 oz. can trap coating (brush in cap) ... Can = $6.25
Item no. PITC32 ... 32 oz. can trap coating ... Can = $15.75

Red Spheres

Use Tangle-Trap Red Sphere Traps to detect and monitor adult apple maggots in your orchard. Early detection allows the scout to take corrective action before the problem gets out of hand. These are *visual* traps only (no lures) and thus reflect the true population in your orchard more accurately.

For best results hang 1-6 spheres per tree, depending upon its size. The lures come with all hanging hardware. The goo needs to be purchased separately. The products discussed here can be obtained from The Green Spot by calling 603/942-8925. Here is the current pricing of these products for 1998...

Item no. PRST ... 1 red sphere trap ... Each = $3.19
Item no. PRST25 ... 25 trap bulk pack ... Case = $53.20

Professional Sweep Net

This product is ideal for scouting field crops. Scouts, you may wish to make a map of your field with a predetermined route marked on it (see Scouting). Following that route, you can slowly cover the area, gently sweeping the net back and forth along the tops of the plants, collecting anything in your path.

Sweep nets like that shown are great for scouting large outdoor areas.

This is an excellent way to collect and

Photos, above and facing page, by M. Cherim
Artwork courtesy of manufacturer

identify the critters on your property — good ones, bad ones, and those in the middle. And on the weekends you can lend the sweep net to the younger people in your life. Let them have fun collecting insects for a school project, hobby, or just to let them loose in the house. Nah. Scratch the last part. Kids don't do stuff like that, anyway.

The construction quality of this professional sweep net is intermediate-duty (see photo, facing page). You don't want the net to be too flimsy, nor do you want it to be too heavy.

This net's features include a 36" sturdy aluminum handle which "feels" right in action, if you know what we mean. The large 15" diameter net mouth, rimmed with flexible but strong flat steel, means you're not going to miss those bugs. And the 27" deep collection bag is made of a tough muslin (it *won't* be confused for a butterfly net).

The product discussed here can be obtained from The Green Spot by calling 603/942-8925. Here is the current pricing of these products for 1998...

Item no. PMPSN ... Medium professional sweep net ... Each = $34.70

Tasco 30x Illuminated Mini-Scope

The world of the bugs is a small one. So small, in fact, that to most of us it goes largely unnoticed. And to the scout, overlooking a few insects could mean the difference between success and failure.

This handy, fit-in-the-palm-of-your-hand illuminated magnifier can change all that. With its large, 7mm. field of view, it offers the scout the uncanny ability to go where the action is; to see the bugs' world in all of its clear, shocking detail. Whether it's on the underside of a leaf, on a flower or bud, or on a sticky trap, this *optically*

Small portable magnifiers are an excellent investment — especially if they are of the illuminated type.

quality-magnifier will take you there. It works great for seeing much of the detail necessary for pest identification. This handy unit will allow the scout to tell the difference between a greenhouse whitefly and sweet potato whitefly — not an easy task without the right equipment (see Figs. 8 & 9, pages 124 & 125).

This rugged, lightweight magnifier features a fully adjustable center focus wheel; a clear acrylic base which allows the scout to accurately pre-position the unit; a textured nonslip finish; and a pre-directed long-life bulb. It also comes with a vinyl, snap-tab carrying case for dust protection.

Please note: The case's snap doesn't usually last long, and the mini-scope bodies are lightweight plastic which, if dropped, *could* break. These mini-scopes, however, are an excellent buy. The optics are really superb (that's why they bear the Tasco name). Every scout should have one. However, if you need something a little more powerful, see the microscopes which follow.

Model # 9704 (see photo, previous page) carries a 1 year manufacturer's limited warranty, included with unit. The unit requires 2 - AA batteries, not included. The product discussed here can be obtained from The Green Spot by calling 603/942-8925. Here is the current pricing of these products for 1998...

Item no. PIMS ... 30x illuminated mini-scope ... Each = $14.95

Ken-A-Vision Microscopes

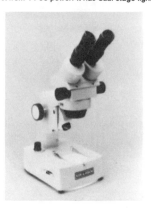

This sophisticated unit (model TT-5Z) can zoom in on an object from 14-90 power. It has dual stage lighting.

Some growers notice they have pests when they find them floating in their morning coffee. While others scout religiously. By the same token, some people happen to notice they have some bugs on their crops and, regardless of what they are, spray (or treat in *some* fashion). While others, though, want to know precisely what they have and how it lives. For the latter, some solutions: the 30x Illuminated Mini-Scope (see previous item) combined with one

Photo courtesy of manufacturer

of the bug identification books (see previous section). And, for those "gotta-know" folks, a Ken-A-Vision stereoscope might be just the ticket.

The Ken-A-Vision line includes 26 variations of 5 models of stereomicro-scopes (dissecting scopes), 26 variations of 7 models of compound micro-scopes, plus a huge selection of other related equipment, including video equipment for showing, viewing or filming your critters.

Price comparisons to the rest of the —scope industry, Ken-A-Vision prod-ucts are priced to sell.

Of special interest to our contacts who, more often than not, tend to look at bugs, are the stereoscopes (see photo, facing page). The products discussed here can be obtained from The Green Spot by calling 603/942-8925. Here is the current price range of these products for 1998...

$160.00 to $850.00 depending upon model and options.
Please call for individual model specifications and exact prices.

getting physical

This has nothing to do with jumping up and down on your pests

Well, maybe it does. After all, jumping up and down on your pests certainly is physical. But, because of the damage this may cause the plants, other methods are recommended first; methods which cause no harm to the plants, but still get the job done. To understand this more clearly, let's first take a good look at the practical actualities of physical methods and what it means.

Getting physical doesn't necessarily mean killing bugs physically or by any other method. It can also include using physical devices or substances which will physically keep bugs away from your crops. This, of course, is a preferred practice — if the pests can be kept away in the first place, they won't have to be dealt with by any other means later. What more could a grower ask for?

For some ideas about products which will keep bugs away, physically, or by building a chemical wall, see the next section, in which both are addressed. And don't forget to use often the cheapest, easiest method of physical pest control: the thumb and forefinger pinch; it's extremely fast and effective (it's 100% effective in controlling the bug(s) in the pinch.

For more detailed information and recommendations concerning barrier products and physical techniques and practices, you are invited to call The Green Spot, Ltd. at 603/942-8925. ⌘

physical means

Diatomaceous Earth

Diatomaceous Earth, commonly called "DE," is an all-natural substance made from the mined and pulverized, fossilized silica shell remnants of unicellular marine algae know as diatoms. Diatoms come in a multitude of shapes and sizes (see illustration, inset).

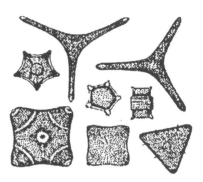

DE has two basic characteristics which make it a valuable tool in the war against bugs: it is microscopically razor-sharp, to crawling things it's like broken glass; it is also a powerful desiccant or drying/anti-caking agent.

DE works extremely well as a barrier. If a pest crosses a line of it, two things will happen: one, the pest's cuticle or skin will be penetrated and; two, the sorptive qualities of DE will rapidly dry the pest out, thus killing it. DE, if kept dry, is an extremely safe and powerful weapon which will retain its qualities indefinitely. Please note, however, DE *must* stay dry to continue working as a desiccant. And its lacerating quality is not enough by itself, unless something else accompanies it: pyrethrum; digestive acids (DE's used to treat intestinal parasites in livestock), etc.

The Green Spot recommends the pure form of DE. The high quality, food-grade type which contains no additives. Its uses are much varied:

Mix it with grain. This serves to control stored grain pests, keep the grain dry, and when ingested by livestock, helps keep intestinal parasites in check. (Use 2% DE, by weight, mixed in feed or grain).

In the home, DE works as an effective barrier against ants (including carpenter ants), cockroaches, silverfish, and more. It can be used in wall voids, cracks, and on carpets and upholstery. Use an old ketchup bottle or similar device for applying it in tight places.

Use it for flea, tick and carpet beetle control in the home: simply sprinkle DE onto your carpets, then take an ordinary household broom, and by gently sweeping back and forth,

Artwork courtesy of Planet Natural

work it into the carpet. There's no need to vacuum it up, and furthermore, once it's in your carpet, you won't succeed in vacuuming it up, anyway. It won't stain, it has no odor, and is completely innocuous to people and pets.

DE is effective in the greenhouse, garden and field. In the greenhouse it is used as a barrier against ants and other crawling pests. This is done by applying it around the perimeter of the structure or on bench legs. In the garden and field, in addition to the greenhouse, it can be used as a general and lethal treatment for a number of different pests. Due to the moisture and humidity levels associated with the uses described above, reapplications will typically be necessary as often as every week, or sooner if it rains (fortunately DE's not very expensive).

This versatile substance can be used in orchards for a multitude of tree pests such as caterpillars and psyllids. Application in trees can be effectively carried out by means of a Dustin-Mizer dust applicator (discussed in the Tools section).

DE can be used as a barrier against crawling pests for the entire home. One way to do this is to apply DE in the space under the siding, next to the foundation. Application in this manner will vary from home to home. Just try to make a line of DE, someplace out of the weather, all the way around your home, between the ground and the wood. Wooden buildings under construction can really benefit from a DE application, where wall voids and such are still accessible. In some cases, if DE doesn't adhere to vertical surfaces, a spray adhesive can be applied before dusting.

Warning: Diatomaceous Earth is absolutely safe to people, pets, etc., when in place. However, when applying DE, irritation of the eyes, nasal passages and lungs is possible because of DE's sorptive qualities. Safety goggles and a dust mask should be worn during application. Once the dust has settled, the potential for irritation no longer exist.

Warning: DE is nonselective. It will kill biological pest control agents just as quickly as it will kill pests. Nontarget organisms will be at risk. Earthworms may also be injured if excessive amounts of DE is used in the soil. Topical applications only are recommended. Use with caution. DE is NOT considered a biorational substance.

The product discussed here can be obtained from The Green Spot by calling 603/942-8925. Here is the current pricing of this product for 1998...

Item no. PDE1 ... DE by the pound ... Lb. = $1.99
Item no. PDE1-20 ... DE by the pound ... 20+ Lbs. = $1.69 / lb.
Item no. PDE50 ... 50 pound bag of DE ... 50 Lb. = $69.50
Item no. PDE50-2 ... 50 pound bag of DE ... 2-3 50 Lb. = $59.50 ea.
Item no. PDE50-4 ... 50 pound bag of DE ... 4+ 50 Lb. = $49.50 ea.

Harvest-Guard Floating Row Covers

Harvest-Guard®
spun-bonded
floating row covers
keep plant pests
away from your
outdoor row crops.
They're safe,
effective and, with
the proper care,
reusable for many
years.

Harvest-Guard, is basically a less expensive version of the well-known *Reemay® fabric. It is packaged in three, ready-to-use sizes: 64"x25', 64"x100' and 128"x25'. The latter is the most popular size for portable hoop supports. *For larger growers we can procure large rolls (67"x2550') of Reemay floating row cover fabric.

In general, floating row covers serve a multitude of purposes in addition to keeping pests away from plants. Some of their attributes are as follows:

☞ They protect tender, young crops from cold weather, frosts and wind; they retain heat, moisture and humidity, and act as season extenders. Please note that heat retention can be a disadvantage in the summer months. However, for many crops, row covers should be removed in the summer months, anyway, to allow for proper pollination and fruit-set.
☞ They allow filtered sunlight, air and water to reach the crops.
☞ By keeping plant pests at bay, disease transmission is reduced.
☞ They will effectively keep many biological pest control agents in with the crop.
☞ Row covers are lightweight, strong, cost-effective and very easy to use. They don't *normally harm plants or stunt their growth — they actually promote fuller growth.
*Abrasion damage can occur in very windy areas.
☞ Their use doesn't significantly add to labor costs.

The products discussed here can be obtained from The Green Spot by calling 603/942-8925. Here is the current pricing of these products for 1998...

Item no. PFRC16 ... 64" x 25' row cover ... Each = $9.95
Item no. PFRC32 ... 128" x 25' row cover ... Each = $13.75
Item no. PFRC64 ... 64" x 100' row cover ... Each = $24.95
Also see Green-Tek's Insecta 500 Row Cover (also discussed in this section).

Artwork courtesy of manufacturer

Row Cover Staples

Use these inexpensive, 5" metal staples to hold floating row covers, plastic mulches and other similar materials securely in place. The product discussed here can be obtained from The Green Spot by calling 603/942-8925. Here is the current pricing of this product for 1998...

Item no. PGS ... 5" staples ... Each = $.20 ea.
Item no. PGSB ... Bundle of 5" staples ... 100 Bundle = $12.50
Item no. PGSC ... Case of 5" staples ... 1000 Case = $85.00

Green-Tek Insect Screens

If it is feasible, we recommend an "open-system." An open system allows insects to come and go as they please. This *may* seem careless. However, most insects are beneficial to growers, not pestiferous. In some cases, though, an open system is simply not possible. For example: a neighboring agricultural operation may negatively affect your own, or your location or crop may invite an unusual number of pests.

When a "closed system" is deemed necessary, screens will make it possible. Before choosing a screen, several factors should be considered, though: sizes available; basic construction; shading value; mesh opening sizes [exclusion ability]; cost; ventilation/cooling reduction; UV stability guarantee; ease of installation; etc.

There are two Green-Tek screens available, of which one may meet your criteria: Green-Tek's Anti-Virus screen and Green-Tek's No Thrips screen. Both are made of transparent, easy to clean, high tensile-strength round polyethylene monofilament. See the specifications below.

Anti-Virus Screen...

Sizes available: Standard widths 43", 78", 118", 137", and custom widths of 59" and 94" available. Cut to desired length. Maximum length 656'.

Basic construction: 50 x 24 ct. mesh; 1/1 weave with a 0.24mm. thread size; 0.037 lbs./yd^2. The open area is 50%.

Shading value: 20%

Light transmission: 80%

Mesh opening sizes: 0.0105"x0.0322", excludes aphids, leafminers, whiteflies, 80% of western flower thrips, and more.

Cost: Very competitive pricing ($.03/ft. below list price min.), see end of section for general rates, call for specifics.

Ventilation/cooling reduction: Varies from structure to structure. Call for more information.

UV stability guarantee: 5 years, prorated.

Ease of installation: Varies from structure to structure. Call for more information.

No Thrips Screen...

Sizes available: Standard widths 39", 78", custom widths available. Cut to desired length. Maximum length 328'.

Basic construction: 81 x 81 ct. mesh; 1/1 weave with a 0.15mm. thread size; 0.024 lbs./yd². The open area is 25%.

Shading value: 33.3%

Light transmission: 66.6%

Mesh opening sizes: 0.0059"x0.0059", excludes all insects mentioned above plus 100% of western flower thrips, and more.

Cost: Very competitive pricing ($.03/ft. below list price min.), see end of section for general rates, call for specifics.

Ventilation/cooling reduction: Varies from structure to structure. Call for more information.

UV stability guarantee: 3 years, prorated.

Ease of installation: Varies from structure to structure. Call for more information.

In summary: Either screen type will reduce the influx of pests, reduce the occurrence of diseases transmitted by pests, and keep biological pest control agents in the greenhouse with the crop.

The products discussed here can be obtained from The Green Spot by calling 603/942-8925. Here is the current pricing of these products for 1998...

> **Item no. PIESAV ... Anti-Virus screen ... Piece = $.63 - $.79 / sq. ft.**
> **Item no. PIESNT ... No Thrips screen ... Piece = $.72 - $.83 / sq. ft.**
> **Call us at 603/942-8925 for a specific quote (have square footage needed).**

Green-Tek's ALUMINET Cool Shading Material

In *some* cases, especially if screening is to be used, shading *may* become even more necessary to reduce greenhouse temperatures. ALUMINET® is a one-of-a-kind shading material; it is a special, knitted screen made from high-density polyethylene strips laminated with a special oxidation-prevention treated aluminum.

The special aluminum coating on Aluminet acts as a mirror, reflecting unwanted solar radiation in the summer and retaining heat in the winter. And, unlike your typical black shade cloth, Aluminet does not absorb heat.

This is really a great product. One customer reported a 10 degree reduction in just one hour of installation!

In addition, the reduction of excessive heat in a greenhouse will increase the efficacy of many biological pest

This photo of the owner's larvae should offer excellent testimony to this product's strength and durability. Have you ever witnessed such abuse?

control agents — plus keep in check the explosive growth of several pests which notoriously get out-of-hand during hot spells.

There are four shading levels of Aluminet available: 30, 40, 50 & 60%. See the specifications below.

ALUMINET 30, 40, 50 & 60%...

Sizes available: Standard widths 6.5', 13', 26', custom widths available. Cut to desired length. Maximum length 1900'

Basic construction: Knitted, oxidation-prevention treated, aluminum laminated, polyethylene strips; 0.135, 0.161, 0.187 and 0.213 oz./ft^2, respectively.

Energy savings: ca 20, 30, 40 and 50% (if used under cover), respectively.

Cost: Very competitive pricing ($.03/ft. below list prices min.), see end of section for general rates, call for specifics.

Tensile strength: 5, 6, 7 and 8 (in accordance with ASTM 191-a), respectively.

UV stability guarantee: 4 years under cover, 3 years outdoors, prorated.

Ease of installation: Varies from structure to structure. Call for more information. Optional mounting fasteners available (see below).

The products discussed here can be obtained from The Green Spot by calling 603/942-8925. Here is the current pricing of these products for 1998...

Item no. PASM30 ... 30% Aluminet ... Piece = $.27 - $.30 / sq. ft.
Item no. PASM40 ... 40% Aluminet ... Piece = $.27 - $.30 / sq. ft.
Item no. PASM50 ... 50% Aluminet ... Piece = $.27 - $.30 / sq. ft.
Item no. PASM60 ... 60% Aluminet ... Piece = $.27 - $.30 / sq. ft.
Call us at 603/942-8925 for a specific quote (have square footage needed).
Item no. PSMMF ... Hinged plastic Aluminet anchoring fasteners ... Each = $.22

Photo by M. Cherim

Green-Tek's Insecta 500 Row Cover / Insect Screen

It's lighter than your typical spun-bonded floating row cover, and therefore less abrasive. It's also stronger. Unfortunately, it is also more expensive. It is, however, because of its special properties, an excellent value. The reason is Insecta 500 is a *woven* material. See the specifications below.

Insecta 500...

Sizes available: Standard width 118" (3 meters), custom widths may be available. Cut to desired length. Maximum length 1000'

Basic construction: Knitted of 100% nylon 6; opening size of 500 microns (0.5x0.5mm.); 0.176 oz./ft² (5 grams); shade value of 65%.

Cost: Very competitive pricing ($.03/ft. below list prices min.), see end of section for general rates, call for specifics.

UV stability guarantee: 3 years, prorated.

The products discussed here can be obtained from The Green Spot by calling 603/942-8925. Here is the current pricing of these products for 1998...

Item no. PI5CRC ... Insecta 500 ... Piece = $.28 - $.30 / sq. ft.
Call us at 603/942-8925 for a specific quote (have square footage needed).

As an authorized Green-Tek distributor, The Green Spot invites you to call us about their other products: Verolite 8mm. Twinwall Polycarbonate greenhouse panels; Green-Lite polycarbonate corrugated sheets; Shade Rite black, green or white knitted polyethylene shade material (30-90%); FVG Sun-Selector greenhouse films; and others. Call 603/942-8925 for samples, info and pricing.

PLEASE NOTE: Green-Tek products are shipped, normally, via UPS ground, F.O.B. Edgerton, WI. Technical data, samples and pricing information available upon request. For screens, we will assist you in checking static pressure compatibility for your greenhouse. Please call us for more information. There are extra charges for most custom cutting, sewing, hemming, custom widths and design work (these services run approximately $.22-.38 per linear foot). Published prices are subject to change, they probably won't, but it does happen, especially with anything made of plastic.

Tree Tanglefoot Pest Barrier

Tree Tanglefoot® Pest Barrier is an extremely sticky, nondrying, resinous compound used to encircle trees in order to effectively stop caterpillars (such as Gypsy moth larvae), ants and other pests from climbing upward. This product may also be used to control ants and other climbing pests on greenhouse bench legs and raised beds, as well.

It is 100% effective, in temperatures of 50-100°F, if applied in a continuous band around the trunk of a tree and kept clean of captured pests and debris.

For certain moths which are unable to fly, such as the female European Gypsy moth — the one that plagues us in the east — Tree Tanglefoot Pest Barrier will effectively exclude them from reaching the tree's upper parts to lay eggs. And thus, we great hunters and gatherers can physically get the eggs down low. For info about doing this, please call us.

Please note that Tree Tanglefoot Pest Barrier will not be effective if the criteria for effectiveness, mentioned above, are not complied with. Which means a weekly inspection of the treated trees (at least) is necessary to assure that the pests are not bridging over the previously-captured. Additionally, if branches of neighboring trees touch the branches of treated trees, crossover may occur. Ballooning, as is part of the Gypsy moth life-cycle, may provide a means of bypassing the barrier, too.

Tree Tanglefoot Pest Barrier comes in a handy tub for brushing or smearing it on, or in a caulking gun cartridge for a more precise applications with less waste. The products discussed here can be obtained from The Green Spot by calling 603/942-8925. Here's the current pricing of these products for 1998...

Item no. PIPB15 ... 15 oz. tub TT pest barrier ... Each = $6.75
Item no. PIPBCT ... 10.5 oz. caulking tube TT pest barrier ... Each = $6.95

Sure-Fire Slug & Snail Copper Barrier Tape

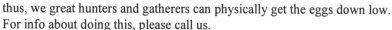

Sure-Fire™ Slug & Snail Copper Barrier Tape is an extremely effective way to keep slugs and snails from getting at your plants. Slugs and snails excrete a slime, upon which they transport themselves.

Thanks in part to this slime, and the copper strip, when slugs and snails come into contact with Sure-Fire Slug

Artwork courtesy of respective manufacturers

& Snail Copper Barrier Tape, an electrochemical reaction occurs and the copper emits a small, electrical charge. This electrical charge is enough to effectively repel these pestiferous mollusks.

This product is 100% percent effective against slugs and snails — they simply cannot cross this copper barrier if it lies vertically in their path.

Sure-Fire Slug & Snail Copper Barrier Tape comes on a 15' roll. It is useful on trees, raised beds, greenhouse bench legs, and anyplace slugs and snails are a problem. It may also be attached to a 2"x4" piece of lumber and flexible edging and placed anywhere as a semipermanent barrier.

The product discussed here can be obtained from The Green Spot by calling 603/942-8925. Here is the current pricing of this product for 1998...

    Item no. PSSCBT ... 1¼" x 15' roll copper barrier tape ... Roll = $6.50

Garlic Barrier Insect Repellent

Garlic Barrier Insect Repellent is made from the oil of a variety of *Allium sativum*... simple garlic, and water. Garlic has been used as a repellent since ancient times, but its odor has kept it out of the commercial greenhouse.

That has all changed now. Garlic Barrier's special formula assures its users of odor-free effectiveness in just 3-5 minutes after application. This has been confirmed at our facility's greenhouse when we used garlic barrier to keep paper wasps at bay. It worked great for us. Recent word from the manufacturer informs us of some new uses for this product — in addition to wasp control: it may be effective as a large and small animal repellent.

Garlic Barrier is 100% safe to humans and beneficials

This one quart bottle will treat up to 119,700 sq. ft.

Photo courtesy of manufacturer

DILUTION RATE

● Mix 1 quart with 3 ounces of horticultural oil or fish oil then dilute 10:1 with water.

1 quart makes 10 quarts of ready-to-use mix and treats up to 119,790 sq. ft.

● Mix 1 gallon with 1 gallon of horticultural oil or fish oil then dilute 100:1 with water.

1 gallon makes 100 gallons of ready-to-use mix and treats up to 20 acres.

NOTE: Dilute only what will be used right away. Concentrate may be stored for up to 3 years.

alike and, according to the manufacturer, a very effective repellent when used as prescribed. This product should be used — like many bio-control agents — before pests arrive. Spray it directly onto your plants every 10 days, beginning early in the season. Use a conventional sprayer, overhead mist system, or even an aircraft if that is what you have.

For best results, spray Garlic Barrier late in the day and mix it with either horticultural oil or fish oil. (We offer horticultural oil. However, we recommend you use fish oil if you're using bio-control agents or if you're certified organic. We can procure fish oil, just give us a call.)

In our opinion Garlic Barrier is very promising. It is very economical, and a little bit goes a long way (see the side-bar). Give it try and tell us what you think; feedback thus far has been mostly positive.

Please note: Garlic Barrier is offensive to bees and wasps. Do not use it if pollination is required or in greenhouses using bumblebees.

The products discussed here can be obtained from The Green Spot by calling 603/942-8925. Here's the current pricing of these products for 1998...

Item no. PGBIR32 ... 32 oz. bottle Garlic Barrier 10:1 conc. ... Qt = $13.95
Item no. PGBIR128 ... 1 gal. Garlic Barrier AG 100:1 super conc. ... Gal = $89.95

This product is also available in drums. EPA flag-word: Caution. REI: 4 hours

Hot Pepper Wax Insect Repellent

We've heard a lot of good things about Hot Pepper Wax; and some not so good. It seems to work very well for about 50% of the folks we hear from, and not so well for the balance. We don't know what to say about this so far

---

1) Not only does this product repel a multitude of plant pests, it also works as a powerful insecticide/miticide. It rivals substances like Avid® miticide and pyrethrum insecticides. And like these products...

2) It is NOT compatible with beneficial insects. This product is nonselective.
But unlike Avid and pyrethrum...
3) This product IS safe to people, animals and the environment. You don't want to get it in your eyes (if you do flush with COLD water), but it IS safe.

Hot sauce or pesticide? Perhaps both?

Hot Pepper Wax™ is an insect repellent, and thus a barrier product. But this compound is so much more. You can also use it as an anti-transpirant, a large and small animal repellent, an insecticide/miticide and, since it leaves a high-gloss shine after application (unlike the film left by most chemicals), it could be considered a leaf shine product — but one with a twist!

The active ingredient in Hot Pepper Wax is also the same natural chemical which puts the bite in south-of-the-border cuisine. That ingredient is capsaicin (cap-SAY-sin). It is common in all solanaceous plants of the pepper plant genus *Capsicum*, or your garden variety jalapeño pepper plant, for example.

Capsaicin disrupts the insects' metabolism and central nervous system. We have absolutely no clue to why this is, but we thought you'd like to know anyway.

Hot Pepper Wax's capsaicin, and the product's other ingredients (culinary herbs including: garlic, onion, mustard, marjoram, oregano, thyme, rosemary and a few others, believe it or not), are micro-encapsulated in a mixture of mineral oil and paraffin.

The paraffin micro-encapsulation is what makes this product easy to handle, produces the leaf shine, smothers pests like a summer oil and protects plants from above-ground, air-transmitted diseases. The paraffin is also what makes this product a long-lasting residual pesticide. Its half-life (a potency reduction of 50%) is roughly 21 days, with complete breakdown in about 30. Maximum product strength is attained in roughly 3 days post-application due

Photo by M. Cherim

to the first sign of paraffin biodegradation. All of these residual periods are affected to some degree by weather, humidity, condensation and dew, and temperature. Example: if it's raining cats and dogs and about a hundred and ten out, it won't last the full 30 days.

The wax also protects from active ingredient rain wash-off (cool water). Yet it is warm-water soluble. That's why you want to use cold water if you get it in your eyes, so you don't breakdown the wax and release the capsaicin.

You can apply Hot Pepper Wax preventively, curatively or for maintenance (we recommend curatively only because of its incompatibility with our good bugs). For insects use a 32:1 mix ratio and usually two to three sprays will do it for a typical season. (For large animal repelling, the mix ratio is 16:1 with at least the same number of applications.) A maintenance application would call for a 64:1. Hot Pepper Wax is not known to cause phytotoxicity.

Again, this product is NOT compatible with beneficials (well, maybe nematodes and *Hypoaspis miles*, the soil-dwelling mite). You should not release the majority of the bio-control agents into an area treated with this product for at least 21 days (this will vary from bug to bug, try your luck if you want). We wanted to carry Hot Pepper Wax because it works, many of you asked us to, and, unlike the *allegedly* "safe" botanical insecticides pyrethrum and rotenone, this product *is* safe — to everything but bugs.

Now that you know about Hot Pepper Wax, you may understand why we say that we had difficulty in placing this product in our manual. We didn't really have reservations, or anything like that, we just didn't know where to put it: with the disease management products? or with last-resort stuff (we *are* recommending only the use of its curative powers to our customers.)? or here in Physical Means section, with barrier products and such? We went with this section because of what the label says. It's a *repellent*.

The product discussed here can be obtained from The Green Spot by calling 603/942-8925. Here is the current pricing of this product for 1998...

Item no. PHPW16 ... 16 oz. bottle Hot Pepper Wax conc. ... Pt = $10.95
Item no. PHPW16C ... Cs 16 oz. Hot Pepper Wax conc. ... 12 / Pt = $78.84
Item no. PHPW32 ... 32 oz. bottle Hot Pepper Wax conc. ... Qt = $19.95
Item no. PHPW32C ... Cs 32 oz. Hot Pepper Wax conc. ... 12 / Qt = $143.64
Item no. PHPW128 ... 1 gal. jug Hot Pepper Wax conc. ... Gal = $75.00
Item no. PHPW640 ... 5 gal. Hot Pepper Wax conc. ... 5 Gal = $360.00

Also available in 55 gallon drums — Call for price. EPA flag-word: Caution. REI: 4 hours.

the biorational way

Just because it comes out of a spray can, it doesn't mean it's bad

Despite the long chemical names, there are a lot of products available today that are considered biorational. This means they are compatible with, or have little impact on, biological organisms. At least that's our meaning; we've heard that people have been tampering with the definitions on some of these new industry words. Go figure.

The beauty of a biorational pesticide is its usefulness for treating a spot flare-up with a short blast or two without sacrificing the entire biological program. Biorational pesticides, if used with care, can be an effective tool to the bio-control practitioner.

Biorational pesticides, for the most part, have fairly short residual periods because they break down quickly. Most are also quite target-selective. Other organisms, other than those for which the product is labelled, are not excessively prone to risk. Care must still be taken though, applying too much of the substance, or spraying it too broadly, may injure nontarget organisms: they could be drowned by the substance, no matter how innocuous.

Botanical pesticides (which, as their name implies, are derived from plants), though touted as earth-friendly, can sometimes be more dangerous than their promoters imply on their snazzy labeling. These should not be confused with biorational products. Carefully read the instructions on the labels of these products. Pyrethrum, rotenone, etc., are really toxic. Their use should be very limited or, better yet, not. Some of these "chemicals" are natural as their makers claim, but, natural or not, they're very powerful chemicals and should be regarded with caution. As far as the two botanical pesticides mentioned being compatible with good bugs: they're not. On the other hand, some botanically derived chemicals, such as Azadirachtin, the active ingredient in neem products, is relatively safe to nontarget organisms. We promote their use when necessary.

Other biorational —cides may include mycoinsecticides, insect specific bacteria, protozoa, etc. For a more complete list of biorational substances, see Fig. 1, pg. 6. If the residual period for a product not listed on our chart is needed, or for more detailed information and recommendations concerning biorational substances, you are invited to call The Green Spot, Ltd. at 603/942-8925.

biorational goods

Azatin XL (Neem)
Biological Insecticide

Azatin™ XL contains Azadirachtin [(AZA) ay-zah-dir-RAK-tin], a neem tree derivative which belongs to the Tetranortriterpenoid chemical family. We consider this neem product to be biorational, as the use of this product in the past three years has not revealed detrimental effects to the good bugs.

The neem tree is famous for its extracts which contain natural phytotoxins, thus having insecticidal qualities (see Fig. 15, this page). Azadirachtin, the number one neem extract, is one the world's oldest known insect growth regulator. Throughout history, Azadirachtin has demonstrated its ability to control over 130 plant pests — 60 of which are common in the United States.

The insecticidal properties of Azadirachtin stem largely from its insect growth regulator (IGR) activity. When ingested or absorbed by a pest insect's larva, Azadirachtin interferes with its metabolism and interrupts the molting process. Mortality typically occurs between successive molts. As such, Azadirachtin may be, perhaps, classified as a larvicide. However, due to its metabolic interference characteristics, Azadirachtin also kills a large number of pest insect pupae [pupacide?]. Adults and eggs, except when used in high concentrations, are not affected. In high concentrations, Azadirachtin will demonstrate its ability as a repellent and anti-feedant.

Insect mortality typically occurs within 3-14 days of ingestion or contact (depending upon pest species, temperature, dosage, weather, etc.). Feeding cessation, though, is prevalent shortly after contact — making the insect

THE NEEM TREE
Fig. 15

The neem tree (Azadirachta indica) is a broad-leaved evergreen which grows abundantly in India, reaching heights of up to 82', but averaging 40-65'. The neem tree, also called the "cornucopia tree," the "Indian lilac," and the "margosa," lives for about 200 years.

By-products of the oil extracted from this highly regarded and special tree include: soap, wax, lubricants, heating fuel and an organic fertilizer — in addition to its popular use as an insecticide and insect growth regulator. By-products of this tree include material for dentifrices, timber and firewood, tannins and feed for livestock.

The neem tree is highly prized, especially the seeds, which contain the most oil, from which Azadirachtin is derived.

The neem tree is just one more example of the riches provided by Mother Nature.

a *pest* no longer.

Most bio-control agents are not be affected by Aza-dirachtin. In fact we know of none which are. However, as a precaution, applications of Azatin XL and similar products, especially high concentration applications, are best made when the significant majority of the good bugs on site are adults. Mammals, birds and other nontarget organisms are typically not affected.

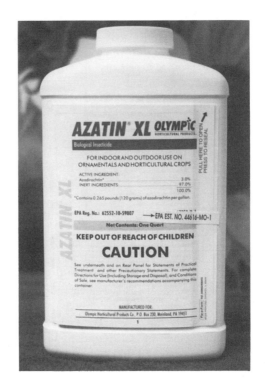

The silver bullet? No way. This stuff's good, though. Use it well.

Insects which are affected the most by Azadirachtin include: *Lepidopteran* larvae, loopers, caterpillars, etc. Others which are affected include most greenhouse pests, confused flour and Japanese beetle larvae, certain turf pests and many more.

Azatin XL is a highly concentrated formula which contains a high percentage of Azadirachtin: 3.0%. Azatin XL, unlike most OTC brands, has no formulation restrictions, contains more active ingredient, is moderately priced, and contains no neem *oil* (which can readily destroy beneficials). Moreover, Azatin XL has a broader range of pest species for which it is listed, and it has a wider window of opportunity in which it can be effective.

Azatin XL is labeled as a primary control of the following pestiferous species: Beet armyworms, cabbage loopers, cutworms, diamondback moth larvae, fall armyworms, fungus gnat larvae, greenhouse whitefly larvae and pupae, Gypsy moth larvae, imported cabbageworms, leafminer species' larvae and pupae, various leafrollers, various loopers, mealybug species' immatures, sawfly larvae, sweet potato and silverleaf whitefly immatures, and webworms. Additionally, Azatin XL is labeled for aphid species, various

Photo by M. Cherim

leafhoppers, and western flower thrips' larvae and pupae, for which it has shown to provide excellent pest suppression. Control has also been demonstrated on cockroaches, Colorado potato beetles, grasshoppers, spruce budworms and tent caterpillars.

Azatin XL is labeled for use indoors and out: greenhouses, nurseries, interiorscapes, fields,

Azatin XL Rate Calculations

TREATMENT AREA	WATER NEEDED	FOR LIGHT PEST PRESSURE USE...	FOR HEAVY PEST PRESSURE USE...
40,000 Sq. Ft.	100 Gal.	10 oz. Azatin XL	16 oz. Azatin XL
20,000 Sq. Ft.	50 Gal.	5 oz. Azatin XL	8 oz. Azatin XL
10,000 Sq. Ft.	25 Gal.	2.5 oz. Azatin XL	4 oz. Azatin XL
5,000 Sq. Ft.	12.5 Gal.	1.2 oz. Azatin XL	2 oz. Azatin XL
2,000 Sq. Ft.	5 Gal.	1 Tbsp. Azatin XL	1 oz. Azatin XL
1,000 Sq. Ft.	2.5 Gal.	0.5 Tbsp. Azatin XL	1 Tbsp. Azatin XL
400 Sq. Ft.	1 Gal.	0.75 tsp. Azatin XL	1 tsp. Azatin XL

trees/orchards, etc. It is listed for well over 105 various bedding plants, flowers, pot plants, foliage and ornamentals (for specifics, please call]. Moreover, Azatin XL is also listed for use in all cruciferous (brassica, cole) crops, all bulb vegetables, all cucurbits, all fruiting vegetables (tomatoes, etc.), fruits and berries, herbs and spices, and all leguminous vegetables.

To the relief of most growers, especially nowadays, applicators of Azatin XL need very little in the way of EPA required personal protective equipment: coveralls or a long-sleeved shirt and long pants, chemical resistant gloves such as barrier laminate or Viton >14ml., shoes plus socks, and protective eye wear. The restricted entry interval (REI) is only 4 hours. An amount of time growers, hopefully, especially retailers, can still deal with.

Additional benefits to Azatin XL include:

1) Low probability of phytotoxicity in treated plants. We've had no negative reports. (However, test plots should treated prior to broad usage.)

2) Azatin XL is known to be compatible with most other chemicals, fertilizers, etc. (call us for specifics).

3) Azatin XL has a great amount of application flexibility; it can be applied as a high or low volume foliar spray, or as a soil drench (for fungus gnats, thrips pupae, etc.).

For best results, as is the case with most pest control products, including our bio-control agents, apply Azatin XL when pest populations are low. Repeat applications every 7 days, or as needed. Use according to the table shown on this page (above, right).

Do not use more than 21 fluid ounces per acre for best results. Also the use of a spreader-sticker will increase its effectiveness. Do not use a spreader-sticker if it is going to be used around beneficials. *Prior* to their introduction it is okay. Try using insecticidal soap or light horticultural oil for this purpose before bio-control introductions to knock down pest numbers (with the soap/oil as an adultacide), then switch to straight Azatin XL and water after introductions. If this product is to be mixed with a spreader-sticker after bio-control introductions have been made, apply it only as a spot treatment spray. In any case, concerning this product, and spray products' use, exercise great caution and restraint. Do not use more than is needed.

The product discussed here can be obtained from The Green Spot by calling 603/942-8925. Here is the current pricing of this product for 1998...

Item no. PAXL32 ... 32 oz. (qt.) Azatin XL ... Bottle = $149.95
Item no. PAXL32C ... Cs 32 oz. (qt.) Azatin XL ... Case of 6 bottles = $809.73

EPA flag-word: Warning. REI: 4 hours.

Dipel 10 G
Sweet Corn Granules
Bt var. *kurstaki* (*Bt-k*)

This strain comes in one formulation: as Sweet Corn Granules. It is used and labeled for the control of the European corn borer and tobacco budworm. Dipel® 10G is a granular formulation which is used as a bait. The target organisms for which this bait is intended find it irresistible, much more than the host plants, in fact. This old, classic variety and formulation is widely used and very popular among growers of corn and tobacco.

Apply directly to corn whorls when 25% of whorls show shot-hole feeding for a 1st generation kill. For a 2nd generation kill, apply when 35% of whorls show feeding. Use 0.5-1oz. per 272 sq. ft. or 5-10 pounds per acre, depending on target pest. Repeat applications as needed.

The product discussed here can be obtained from The Green Spot by calling 603/942-8925. Here is the current pricing of this product for 1998...

Item no. PBTK8 ... 8 oz. can Dipel 10 G ... Each = $6.25
Item no. PBTK1 ... 1 lb. can Dipel 10 G ... Each = $9.75

EPA flag-word: Caution. REI: 4 hours.

Dipel 150
Dust
Dipel .86 WP
Wettable Powder
Bt
Liquid Concentrate
Bt Berliner (*Bt*B)

This strain comes in 3 different formulas: dust, wettable powder & liquid concentrate (see photo, inset).

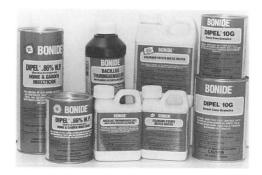

Universally used to combat pests, Bt comes in myriad shapes and sizes.

The **Dipel®** 150 dust is used and labeled for the control of cabbage loopers, imported cabbageworms (see photo, facing page) grape leafrollers, and tobacco and tomato hornworms.

Apply when the pests are first noticed. Use 0.25-4ozs. per 435 sq. ft. or 12 -50 pounds per acre, depending on the target pest. Repeat applications every 4-10 days, as needed.

Dipel® .86W.P. wettable powder is used and labeled for the control of cabbage loopers, diamondback moth larvae, Gypsy moth larvae, elm span-worms, inch worms, leaf folders, sod webworms, tobacco and tomato hornworms, and tomato fruitworms.

Apply when the pests are first noticed. Use 1-8 tablespoons per gallon of water. Use 2-10 gallons per acre. Rates vary according to crop and target pest. Repeat applications every 4-10 days, as needed.

BtB Liq. Conc. liquid concentrate is used and labeled for the control of alfalfa caterpillars, cabbage loopers, fruit tree rollers, grape leafrollers, Gypsy moth larvae, imported cabbage worms, omnivorous leafrollers, orange dogs, red-humped caterpillars, rind worm complex, and tobacco and tomato hornworms.

Apply when the pests are first noticed. Use 1-4 tablespoons per gallon of water. 1oz. treats 680 sq. ft., or 0.25-2 qts. per acre, depending on the target pest. Repeat applications every 7-14 days, as needed. WARNING: PRODUCT CONTAINS OIL.

Photos, above and facing page, by M. Cherim

The products discussed here can be obtained from The Green Spot by calling 603/942-8925. Here is the current pricing of these products for 1998...

Item no. PBTBD1
1 lb. can Dipel 150 D
Can = $6.35

Item no. PBTBD4
4 lb. bag Dipel 150 D
Bag = $17.50

Item no. PBTBW4
4 oz. can Dipel .86 WP
Can = $6.50

Item no. PBTBL8
8 oz. bottle Bt liq. conc.
Bottle = $8.35

Item no. PBTBL16
16 oz. bottle Bt liq. conc.
Bottle = $11.95

EPA flag-word: Caution.
REI: 4 hours.

About Bt   Fig. 16

Bacillus thuringiensis [(Bt) BAH-sil-us thur-inn-gee-en-SIS] is a pathogenic, bacterium-based, microbial insecticide which targets a specific individual or group of organisms.

Bt is named after the German town of Thuringia where it was discovered in 1911 in diseased flour moths. Since that time, 35 other strains have been identified. Each strain attacks specific target organisms. Commercial marketing of Bt began in 1958 with variety kurstaki.

Bt, a substance consisting of spores and protein crystals, is a stomach poison which must be ingested by the target organism in order to be effective. Upon ingestion, the spore, which comes into contact with the highly basic (alkaline pH>7) gastric liquids and potent enzymes inside the stomach of the pest, begins to breakdown. When the spore breaks down, it exposes the protein crystal. The crystal neutralizes the enzymes which serve to protect the stomach lining of the pest from its own digestive juices. When the protective enzymes are neutralized, the digestive juices begin to eat holes through the target organism's stomach wall. The stomach's contents enter the affected organism's body cavity and blood stream. At this point, all feeding stops— the target organism is no longer a destructive pest. This poisoning of the blood, known as septicemia, may also cause paralysis. The cessation of feeding happens in as little as a few hours post-ingestion, and biolysis or dissolution of the pest will occur in just a few days —it gets yucky. After death, the bacteria inside the host will continue to parthenogenetically reproduce. However, new spores and toxic crystals are rarely produced in the subsequent generations —Bt is a onetime killer.

The primary, and most popular, target organisms for which Bt is used, are the destructive larvae of several moths and butterflies: caterpillars. Caterpillars affected by Bt will normally darken and hang downward at an angle of 90° from the branch to which their prolegs (pseudolegs used to support and aid mobility to their massive larval bodies) are attached.

Bt is very specific to the pest for which it is intended and labeled. In laboratory tests, toxicity in mammals and nontarget organisms is virtually nil. Therefore, Bt's use in a bio-control or IPM program is highly recommended if the target pest, for which the formula is labeled, is one requiring control. Please note, however, that some formulations— those with a petroleum base or other adjuvant used as a spreader-sticker or surfactant, etc.— may be lethal to nontarget organisms such as some beneficials. Caution should be exercised if the possibility of physical contact with nontarget organisms exists.

Cabbage worms on broccoli leaf.

Colorado Potato Beetle Beater
Bt var. *san diego (Bt-sd)*

This strain comes in one formulation: a liquid concentrate. It is used and labeled for the control of Colorado potato beetle larvae and elm leaf beetle larvae and adults.

Apply at the very first sign of larval presence. Use 1.5ozs. per gallon of water. 1oz. treats 1000 sq. ft. Repeat applications every 4-10 days, as needed.

The product discussed here can be obtained from The Green Spot by calling 603/942-8925. Here is the current pricing of this product for 1998...

Item no. PBTSD8 ... 8 oz. bottle Bt san diego liq. conc. ... Bottle = $9.95
Item no. PBTSD16 ... 16 oz. bottle Bt san diego liq. conc. ... Btl = $15.35

EPA flag-word: Caution. REI: 4 hours.

Mosquito Dunks
Bt Berliner var. *israelensis (Bt*B-*i)*

There is one stable formulation of this variety: a compressed briquette - Mosquito Dunks®. Mosquito Dunks are used and labeled for the control of mosquito larvae. These *Bt*B-*i* briquettes may be used anywhere there is standing, stagnant water: puddles, birdbaths, swampy areas, etc.

These briquettes can even be used in vernal pools (pools, puddles, etc., which dry out between rainstorms — mosquitoes exploit these sites, too. Mosquitoes lay eggs in vernal pools, and when it rains, the eggs hatch. Mosquito Dunks work in vernal pools because they remain effective indefi- nitely when dry. However, as soon as they get wet, they are reactivated. If they dry out again, they can be stored again, and that cycle may be continued until the briquette is completely dissolved.

This is easy-to-use product which is very effective. For even better mosquito control, also try enticing bats to live on your property. (Bat houses are covered in this manual.)

Apply 1 briquette per 100 sq. ft. of standing water, regardless of its depth. Briquettes may be broken into pieces for smaller pools. Repeat applications every 4 weeks or so, as needed or when briquettes dissolve completely.

The product discussed here can be obtained from The Green Spot by calling

603/942-8925. Here is the current pricing of this product for 1998...

Item no. PBTID6 ... 6 pk. Mosquito Dunks ... Each = $10.50
Item no. PBTID20 ... 20 pk. Mosquito Dunks ... Each = $25.25

EPA flag-word: Caution.

Nosema locustae

Nosema locustae (noh-SEEM-ah lo-KUS-tay-ay) is an important pathogenic, protozoan insecticide used and labeled as a control of 58 grasshopper species and the Mormon cricket. It will also suppress the black field cricket and a species of pygmy locust. It affects no other living creatures.

N. locustae is a disease-carrying spore which must be ingested by the target organism. To promote ingestion, *N. locustae* spores are mixed with a flaky wheat bran which grasshoppers find irresistible — more so, in fact, than the protected crop. The mixture consists of approximately 1 billion microscopic spores per pound of bran.

Once ingested, *N. locustae* spores breakdown as soon as they reach the grasshopper's midgut. When the spores break down, the disease, through the normal processes of nutrient absorption, enter the grasshopper's circulatory system and is passed through the body, ultimately ending up in the grasshopper's lipids or fat cells. There the disease blossoms, disrupting circulatory, excretory and reproductive functions. The grasshopper's disfigurement and/or death will ensue.

50% of the grasshoppers who ingest this tainted bran will die from the disease. The other 50% will pass the disease on to subsequent generations; they, too, will be infected and either pass the infection on or die. 100% of those which consume the bran will become infected, and thus show a marked reduction in feeding.

Another way *N. locustae* is effective is through the grasshoppers' willingness to partake in cannibalism — which further spreads the disease.

If you see 8 grasshoppers (see photo, next page) per square yard, your crops may be endanger, at least according to the USDA. If you have such a population, you should apply one of the two brands we offer. Which, depends upon the size of the treatment area. If grasshopper populations are excessively high, i.e., greater than 40 per square yard, 1-2 repeated applications in the same season should be considered. In fact, since our trials of this

A damaging grasshopper species on a sunflower bud.

product in the past two years, we have to recommend that everyone try multiple applications. We no longer feel, contrary to manufacturers' recommendations, that a single shot is enough. Much will depend on the species you have, too. We don't know what works best with what and ask that you experiment — then tell *us* the right dosage to use.

Grasshoppers do not undergo complete metamorphosis (development process): they molt through 6 nymphal instars (stages), and do not have a larval and pupal stage as do many insects. The most desirable time to apply *N. locustae* and its carrier bait is during the early stages: 2nd and 3rd instars (when grasshoppers are 3/8" to 1/2" long). The best areas to apply this tainted bait is in south-facing areas with sandy soil. These are good breeding/hatching grounds, and young 2nd and 3rd instar grasshoppers will be present in high numbers.

Application methods vary according to the size of the area to be treated. They range from hand broadcasting, to hand-held spreaders, to ATV hopper/spreaders, to trucks, tractors and even aircraft.

Apply bait in the morning when temperatures are above 60°F and no rain is forecasted so the grasshoppers are in the eating mode, and have time to do exactly that — rain will make the bran unpalatable and ineffective. Spread the bran at a consistency of no less than 25 flakes per square foot. 1 pound of bran bait will treat 1 acre. Grasshoppers will eat the bait quickly.

We offer two different brands of *N. locustae*: Nolo Bait® and Semaspore™. Both are equally effective. Nolo Bait comes in a large 50 pound size; Semaspore comes in a 1 and 5 pound sizes. Neither can be stored for more than three months, so order only what you need. During the off season (other than spring and summer) we may not have this product in stock. Advance

Photo by M. Cherim

ordering, during any time of year, is recommended. Semaspore is registered for use in every state. Nolo Bait, however, is registered for use only in certain states: AZ, CA, CO, FL, ID, KS, MI, MN, MO, MT, NE, NV, NM, ND, OH, OK, OR, SD, TX, UT, VT, VA, WA & WY. (Confirm prior to ordering.)

The products discussed here can be obtained from The Green Spot by calling 603/942-8925. Here is the current pricing of these products for 1998...

Item no. PNLS1 ... 1 lb. can Semaspore ... Can = $13.95
Item no. PNLS5 ... 5 lb. bag Semaspore ... Bag = $29.95
Item no. PNLN50 ... 50 lb. bag Nolo Bait ... Bag = $99.95
Item no. PNLN50-2 ... 50 lb. bag Nolo Bait ... 4+ Bags = $74.96 ea.

EPA flag-word: Caution.

Beauveria bassiana

We've heard some pretty good things about these products, and very little bad (we'll cover that, read on). And we are pretty comfortable in saying that these products are fairly compatible with beneficials.

Beauveria bassiana (BOH-værr-ee-ah bah-SEE-ohn-nah) mycoinsecticidal products, such as Naturalis-O and BotaniGard, contain viable spores of this naturally occurring insect-specific pathogenic fungus. Naturalis-O also contains a vegetable oil emulsifier. BotaniGard, according to the manufacturer, contains *many* more viable spores than the competitive product, but it has a petroleum oil base. The latter is currently the only product labeled for use on vegetable crops and may soon have a vegetable oil base, too.

The *B. bassiana* fungus works by physically coming into contact with the host insect, one of myriad species of sucking and chewing pests affected by the fungus. This is accomplished by spraying the viable fungal spores directly onto the insect, using conventional spray equipment, or if the insect travels through the sprayed area while the leaves still contain viable fungal spores, or through insect to insect contact (horizontal or lateral translocation).

Once fungus-insect contact is established, the spores attach themselves to the insect's cuticle (skin or covering). After securely attached, the spores secrete specific enzymes which dissolve the cuticle at the point of contact. This facilitates the entry of the fungus's hyphae (kind of like roots) into the body of the pest.

Like that of roots, the hyphae absorb and transport nutrients to the spores, which grow and multiply, attacking the pest to an even greater degree, sending out more hyphae. The theft of the moisture and the robbing of nutrients causes the pest to die. Depending upon the penetration resistance encountered by the fungal spores, which varies from pest to pest, death, or least a detectable decrease in activity, can occur in as little as 24 hours — amazingly quick for a *biological* control. The average maximum amount of time is 5 days.

The adhesion and growth of the fungi can sometimes be detected by the scout. Usually there will be *some* sign: cuticle color change, lack of insect mobility, lack of feeding, no-nooky, etc. In *some* cases, after the death of the target insect, the fungal growth may actually be seen on the cuticle of the dead pest. This will appear as a whitish "fuzz." This is not always visible, though, and cannot be relied upon to serve as an indicator of product effectiveness.

Effective death-from-within of pests can affect all life stages — on contact — these points represent two advantages over *Bacillus thuringiensis*. However, as usual, there are some drawbacks, too: 1) These products may have compatibility problems with some fungicides (you should wait 48 hours before applying fungicides), which makes sense, since this *is* a fungus we're dealing with; 2) This product, we've heard, needs to stay wet on the foliage as long as possible a achieve maximum results, this, of course is conducive to fungal disease growth, but we're not sure how valid this claim is. According to one label, residual effects can last 3-7 days post-application; 3) *B. bassiana* may be pathogenic to honeybees, and thus incompatible, (though compatibility is suspected with *Bombus* species of *bumble*bees; 4) Nozzle screens of finer than 25 mesh should be removed to prevent clogging; 5) Phytotoxicity may occur. It is advisable to treat a test plot before spraying the entire range; 6) You shouldn't use this product on poinsettias after the development of their bracts.

Some of these products' advantages, not mentioned previously, include: its short 4 hour REI, same as *Bt*; their ability to manage resistant pests without the fear of them becoming resistant to them, as is the case with beneficials; their long shelf life (2 years from date at 35-78°F*) and; lastly, they can effectively control many economically important pests at an acceptable price. *(Please note, prolonged storage at 90°F+ may impair product viability. Use within 8 hours once diluted.)

Pests affected include all stages of whiteflies, thrips, fungus gnats, aphids, shore flies, mealybugs, leaf hoppers, strawberry and obscure root weevils, lygus bugs, flea and cucumber beetles, black vine weevils, red and two-spotted spider mites, psyllids and many other pests, including several species of destructive leaf-eating moth larvae.

The products discussed here, at press time, can NOT be obtained from The Green Spot. The Green Spot has been disallowed to offer Naturalis-O. It is unclear to The Green Spot's management as to the actual reason(s); perhaps it is a new, bizarre way of doing business?!? If you, as a disappointed client, would like to have us offer Naturalis-O, or would like to try and buy it direct, please call Troy Biosciences, Inc. at (602)233-9047. Mycotech Corp., the makers of BotaniGard, (406)782-2386, is still considering us as a distributor. For an update, please give us a call and we'll give you the latest information as we know it.

Sorry, pricing not available at press time.
Please call 603/942-8925 for availability and pricing.

EPA flag-word: Caution. REI: 4 hours.

Milky Spore Disease
Bacillus popilliae

In the 1950s the USDA employed the use of a newly discovered natural pest control product developed by Dr. Samson Robert Dutky in and around the Washington, DC area — including the White House lawns. The product was called Milky Spore Disease, scientifically known as *Bacillus popilliae*. The product was discovered to be very effective in controlling the larval grubs of the Japanese beetle, *Popillia japonica*.

Since that time, many bacterial products containing this spore have been produced under various trade names. In the early 1990s, due to certain unusual circumstances, *B. popilliae* was largely unavailable. Now, however, it's back in full force — the original manufacturer is at it again.

Here are some facts concerning the use of this product:

1) It is a useful and effective control of Japanese beetles, but can take a considerable amount of time to effectively control large numbers of pests, even though it starts working on the feeding grubs as soon as it enters their

The 10 and 40 oz. cans are shown. Also shown is the handy dispensing tube.

realm. It's wise to start a Japanese beetle control program with parasitic nematodes (see soil pest controls) or other control first. As always, we cannot recommend the use of Japanese beetle traps (they may attract both mating pairs but may not capture them all).

2) Control, once achieved, can last an incredibly long time. 20-30 years of unsupervised control is not uncommon. This is true even in the coldest areas of the contiguous states; freezing will not negate this product's effectiveness. If the grubs are controlled, the spores will lie dormant in the soil until more host material becomes available. Typically, only one application is needed for the life of the lawn, unless heavy rains wash the product away before it has a chance to work its way into the soil. Another factor which may necessitate an additional application would be if the product has leached too deeply into the soil — below the soil strata just under the sod where the beetles' larvae hang out before pupating and emerging as adults.

3) *B. popilliae* can be used anywhere there are Japanese beetle grubs: vegetable gardens, lawns, even golf courses. Even though it safe to use just about anywhere, it is not recommended for use on pasture lands.

During the spring and fall, when the grubs are present in the soil and feeding, apply one level teaspoon of Milky Spore Disease powder in spots every three

Photo by M. Cherim

feet, in rows three feet apart — creating a grid-like network. This product can be applied to the surface of the lawn, then watered in (it is not effective until it reaches the root zone). We've heard that more closely located "spots" or digging it in is required under certain circumstances, but the manufacturer says this is not so.

The 10 oz. container treats 2500 sq. ft. of surface; the 40 oz. (2.5 lbs.) unit treats 10,000 sq. ft.; use 10 pounds per acre. To make application and dosing a lot faster and easier, get the dispenser tube. For large scale applications 50 and 100 pound units are available; they treat 5 and 10 acres, respectively.

The products discussed here can be obtained from The Green Spot by calling 603/942-8925. Here is the current pricing of these products for 1998...

Item no. PMSDT ... Dispenser tube ... Each = $7.99
Item no. PMSD10 ... 10 oz. Milky Spore Disease ... Can = $29.95
Item no. PMSD40 ... 40 oz. Milky Spore Disease ... Can = $89.95
Item no. PMSD50 ... 50 lb. Milky Spore Disease ... Drum = $1539.00
Item no. PMSD100 ... 100 lb. Milky Spore Disease ... Drum = $3050.00

EPA flag-word: Caution.

plant diseases

As it is with humans, plants can get sick too

Illness breeds illness. Plant diseases can make plants more susceptible to disease. Moreover, it can also make plants much more prone to pests. Ironically enough, many diseases are transmitted by pests to begin with. Oh what a vicious cycle it can be.

Diseases in plants are typically bacterial or fungal. Thus, their controls are either bactericides or fungicides. And there are many —cides available out there, all capable of controlling a specific disease and list thereof. Not all are safe to beneficials though, which is the reason this section was thought of in the first place. In the section which follows, there are many products which are the safest to use. If it's not the actual compound that's safe, then it is when the product is applied which makes it so.

As with biological pest controls, prevention is the answer to ultimate disease control: stop it from happening and you won't have to deal with the consequences. Most of the "safe" products shown in the next section work best as preventive medicine for plants; though some work in both capacities [as curatives too].

Let's back up a moment before we become nozzle-heads. There are some simple practices which may help prevent diseases from occurring — even without the preventive dusts and sprays:

1) Plant spacing is very important. The greater the distance between plants, the less likely that spores and such will pass from one plant to the other. Moreover, gaps will increase air circulation and light to surface areas. This will keep plants dryer. As is the case with most diseases, requirements for optimum manifestation will include darkness, heat and, especially, moisture or high humidity.

2) Insect pests are probably the number one cause of disease spread among plants. Good insect control is really the key to preventing many disease problems. Insect screens are a terrific idea when considering pest control from the disease standpoint; like a condom, screens provide maximum separation between the disease and plant.

3) Scouting your plants will usually provide early insights to disease manifestation. Finding disease symptoms on one plant — and removing that plant — can really

save the day. Remember to wash your hands and the general area of the infected plant immediately after disposal. This is very important because...

4) Sanitation is fundamental. Like some pests, we can transmit plant diseases, too. Entire greenhouse sanitation procedures can be of immense value to growers. Clean up now, avoid headaches later. This may also include the sanitation or sterilization of your preferred media. Many plant pathogens get their start in tainted media. Nowadays, though, you can often add living organisms to the soil which will do the dirty work for you — naturally, without sometimes costly sterilization procedures. Two such products are discussed in the next section.

5) Outdoor crops should be rotated often. As many growers know, you don't want to plant cole crops in areas where another cole crop had been grown within the past three years. Like plants can sometimes increase the likelihood of problems associated to that particular family or variety.

6) Get more information. There are books available which will help you identify, prevent and correct disease problems (some are listed in this book). A good place to start is with your local cooperative extension office; they may be familiar with the common plant diseases which tend to be problematic in your given area.

7) Scout. Yeah, we're back to that again. Just can't stress enough its importance.

For more detailed information and recommendations concerning plant diseases and ways to effectively deal with them, you are invited to call The Green Spot, Ltd. at 603/942-8925.

disease controls

Actinovate - Soil & Plant Inoculant

Like Mycostop, discussed next, this product contains a beneficial organism. The species of organism differs, though. Actinovate™ contains 10^6-10^9 colony forming spores of another *Streptomyces* species: *S. lydicus*, which is, specifically, a saprophytic rhizosphere colonizing actinomycete which was isolated in and separated from a linseed plant in an area of unusually high natural suppression of soil pathogens.

In warm and moist conditions, the spores germinate and form mycelia which attach themselves to the plant's roots. In these conditions, reproduction runs rampant and, before long, the entire root system of the plant is covered with protective mycelia.

The mycelia live in a symbiotic or co-dependent relationship: the plants make a home for the mycelia (their tenants); and the mycelia convert minerals and nutrients into food for the plant, protect the plant from harmful soil pathogens (both promote growth), and promise to keep the stereo turned down after ten-o'clock on weeknights.

Unlike Mycostop, Actinovate is not a fungicide, it is simply a soil inoculant. However, it *does* display some fungicidal qualities, such as keeping several economically important soil diseases at bay. This is not done by killing them, it is merely done by out-competing them for the livable space amongst the plants' roots by producing powerful antifungal metabolites. Technically, it can't be a fungi*cide* because it doesn't directly kill.

However, some pathogens which are held at some level of control include *Pythium* spp., *Fusarium* spp., *Phytophthora* spp., *Sclerotinia* spp., and one species from each of the genera *Rhizoctonia, Phanerochaete, Coriolus, Postia, Caldariomyces, Gloephyllum, Geotrichum* and *Verticillium*.

The advantages of this product may be obvious, it seems to be economical (if it works as claimed by the manufacturer, which it seems to do based on numerous university tests), it protects plants and promotes heathier, fuller growth in a shorter period of time, which in the case of a some food crops, will boost yields.

This is a product we suggest you try. It comes in three handy formulas, and in quantities easier to deal with than typical products of this sort...

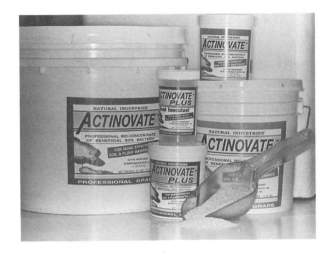

With a multitude of sizes and formulations, you can address your needs with more precision than with similar products.

Actinovate Granular 10[6] should be incorporated into, or used to top-dress beds, planters, potted plants, and soilless mixes. It can also be used to top-dress row and field crops, and turf-grass. 1 pound of granular treats: 1000-2500 sq. ft. of turf, depending upon the severity of the problem; 1 cubic yard of soil or mix; or 360 sq. ft. of bed prep. In containers use 1/2 tsp. per 4" pot, 1 tsp. per 7" pot or 1 tbsp per 1 gallon container.

Actinovate Plus Wettable Powder 10[8] can be used in turf, trees and orchards, field transplants and surface applications. It can also be used to pretreat seed potatoes, and cotton and rice seed. We were informed in 1997 that this product did not breakdown fully and caused the clogging of drip irrigation lines. The manufacturer has informed us that this problem has been addressed and should no longer be of concern. Use 8-16 oz. mixed with sufficient water, per acre of turf, or mix 8 oz. with 300 gallons of water to treat up to 9600 field transplants, or use 4 oz. with 50 gallons of water for tray dipping. In orchards, use 16 oz. with 250 gallons of water and inject 5 gallons into the root-zone of each tree. For seed treating use 2-3 oz. per 50 pounds of seed. Or use the seed coating, next...

Actinovate Seed Coating 10[9] can be used for most seed, including vegetable seed. Use 4 grams per pound of seed.

Photo by M. Cherim

We are not aware of any of these products' limitations. As far as we can determine there are no restrictions to use, storage or shipability. Use some of the recommendations for Mycostop, they may apply. Try it, tell us how well you like it — or don't.

The products discussed here can be obtained from The Green Spot by calling 603/942-8925. Here is the current pricing of these products for 1998...

Item no. PAIG1 ... 1 lb. Actinovate granular ... Jar = $16.75
Item no. PAIG6 ... 6 lbs. Actinovate granular ... Tub = $70.50
Item no. PAIG25 ... 25 lbs. Actinovate granular ... Bkt = $217.25
Item no. PAIWP16 ... 16 ozs. Actinovate wettable powder ... Jar = $149.50
Item no. PAIWP25 ... 25 lbs. Actinovate wettable powder ... Bkt = $2725.00
Item no. PAIWP40 ... 40 lbs. Actinovate wettable powder ... Bulk = $4185.00
Item no. PAIST120 ... 120 grams Actinovate seed treatment ... Jar = $270.00

Mycostop
Biofungicide

The active ingredient in MYCOSTOP® is a naturally occurring, soilborne bacterium: *Streptomyces griseoviridis,* Strain K61 [10^8 cfu (colony forming units) per gram of product]. Mycostop contains 30% dried spores and mycelium of ray fungi.

Mycostop is mass produced by fermentation, then formulated as a wettable powder. It is sealed from air and moisture in small, hermetically-sealed packets. The packets should not be opened until ready for use; if prematurely opened, Mycostop will lose its potency. The packets may be stored prior to use in a cool, dry place, in a refrigerator or freezer.

Mycostop works in three ways: one, is by out-competing several undesirable pathogenic fungi for their living space and nutrients by colonizing plant roots more expeditiously than other microorganisms. The second, is by excreting various enzymes and metabolites which inhibit the growth of the undesirable pathogenic fungi. And lastly, *Streptomyces* sp. bacteria are known to spur growth in their host plants, making them more vigorous, healthy and strong; this makes the plants treated with Mycostop less susceptible to pathogens — especially in stress-promoting growing conditions. Due to this ability to increase plant vigor and general health, Mycostop has shown researchers in trial after trial dramatic yield increases.

Mycostop, once activated by coming into contact with a humid growing medium, stays active in a wide range of conditions. Optimum substrate temperatures are from 50-77°F, activity *limits* are from 41-113°F. (These may be the same tolerances for Actinovate, see previous). Mycostop is active in mediums with a pH value of 4-9. It works well in organic and inorganic mixes.

The refrigerated shelf-life of this product has been greatly increased.

Adequate moisture, enough to keep in-medium humidity levels high, will increase the activity level of Mycostop.

The main target pathogens of Mycostop are *Fusarium* spp. fungi which cause wilt, basal and root rot diseases in greenhouse cultivated fruits, herbs, ornamentals and vegetables. Mycostop also controls seed and soil diseases such as damp-off, root rot and foot rot caused by *Alternaria brassicicola* and/ or *Phomopsis* spp. Additionally, suppression of *Botrytis* gray mold, and *Pythium* spp. and *Phytophthora* spp. root rots is possible.

Mycostop is labeled and used to prevent the pathogens specified above. As a seed treatment, it is labeled for use with several crops including *all* cruciferous crops (brassica or cole: broccoli, cabbage, etc.), *all* leguminous crops (peas, beans, etc.), and *all* root crops (carrots, beets, etc.). As a soil spray or drench, it is labeled for wide range of different crops: cruciferous, cucurbits, leafy greens, peppers and tomatoes, just to name a few.

As a seed treatment: Mix 2-8 grams of Mycostop per kilogram of seed.
As a soil drench: Mix 5 grams of Mycostop per every 13 gallons of water.
As a soil spray: Mix 5 grams of Mycostop per every 1.3 gallons of water.
As a dip treatment for cuttings (ornamentals): Use a 0.01% Mycostop and water solution. Note: IBA (indole-3-butyric acid) and/or NAA (1-naphthaleneacetic acid) may be used in the same dip treatment.

For tomato growers: try using 1 - 5 gram pack per every 1000-1200 plants every 6 weeks. For seedlings 1 gram will treat 108 sq. ft. These rates seem to

Photo by M. Cherim

work well. They are certainly more economical than if used as suggested above. Try these rates on other crops, as well; satisfactory and economical results may be uncovered there, too.

We feel, that in addition to sound cultural practices, Mycostop is the *best* first line of defense in the prevention of greenhouse crop diseases. Mycostop is useful as a treatment for cuttings, seeds and growing mediums, and it is compatible with most chemical fertilizers, pesticides and acaricides (miticides), fungicides, rooting hormones and growth regulators. It is also compatible with parasitic nematodes and other bio-control agents. We've experienced no problems with the flexibility of this product.

The product discussed here can be obtained from The Green Spot by calling 603/942-8925. Here is the current pricing of this product for 1998...

Item no. PMBF1 ... 1 gram pkt. Mycostop ... Pkt = $10.50
Item no. PMBF5 ... 5 gram pkt. Mycostop ... Pkt = $33.50
Item no. PMBF25 ... 25 gram pkt. Mycostop ... Pkt = $159.50

EPA Flag-word: Caution. REI: 4 hours.

Copper-Sulfate Fungicide

Copper-Sulfate Spray or Dust Copper Bordeaux substitute is an organic fungicide containing 7% copper sulfate (metallic). It is labeled for use on eleven vegetable crops, six fruit crops, and a wide range of others including deciduous and coniferous trees, flowering shrubs and ornamental plants and roses. It is effective in preventing a wide range of various blights, spots, certain rots, downy and powdery mildew, leaf blister, anthracnose, scab, stem canker, Septoria and Stemphylium leaf molds and more.

Copper-Sulfate may be mixed with water and sprayed, or applied as a dust. It has no insecticidal qualities, and will not burn plants. We've had no incompatibility problems with this product. Copper products, versus other fungicides' ingredients, seem to be the safest on the good bugs. (Also see Phyton 27, also in this section.)

In most instances, very early application is required, and must be repeated. As a spray: mix 0.5-6ozs. per gallon of water. Call for crop- and disease-specific application information.

The product discussed here can be obtained from The Green Spot by calling 603/942-8925. Here is the current pricing of this product for 1998...

Item no. PCSB1 ... 1 lb. shaker Copper-Sulfate (spray or dust) ... Can = $6.75
Item no. PCSB4 ... 4 lb. bag Copper-Sulfate (spray or dust) ... Bag = $12.50

EPA Flag-word: Caution.

Liquid Copper 4E

Liquid Copper 4E is an organic fungicide containing 48% copper sulfate (4% as metallic, 44% as salts). It is labeled for use on eight vegetable crops, ten fruit crops, and a wide range of others including coniferous trees, and roses. It is effective in preventing several blights, spots, rusts, certain rots, downy and powdery mildew, leaf blister, anthracnose, leaf curl, shothole and more.

Liquid Copper 4E has no insecticidal qualities, and will not burn *most* plants; certain grape varieties may suffer some marginal leaf burn, however.

In most instances, very early application is required, and must be repeated. Late season application is required for some crops. Mix 2-10 tsp. per gallon of water. Call for crop- and disease-specific application information.

The product discussed here can be obtained from The Green Spot by calling 603/942-8925. Here is the current pricing of this product for 1998...

Item no. PLCF16 ... 16 oz. Liq. Copper fung. conc. ... Btl = $9.65

EPA Flag-word: Caution.

Micronized Sulfur Fungicide

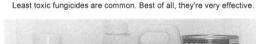

Least toxic fungicides are common. Best of all, they're very effective.

Micronized Sulfur Fungicide is an organic fungicide which can be dusted or sprayed. It contains 95% micronized (ground to 4-5 microns providing excellent coverage) elemental sulfur. It is labeled for use on two vegetable crops, fourteen fruit crops,

Photo by M. Cherim

including *all* citrus, and a wide range of others including ornamental shrubs, perennial plants, and roses. It is effective in preventing several spots, rusts, certain rots, powdery mildew, blights, blotches, frog eye, shothole, scab and more.

Micronized Sulfur Fungicide also has insecticidal qualities. It is effective on certain crops against: soft black, citricola, purple, red, and yellow scale insects; plus citrus red, European red, rust, and two-spotted spider mites; and thrips. Micronized sulfur may harm certain apple, grape and pear varieties.

This product may also be used to lower soil pH values (make more acidic: pH<7).

In most instances, very early application is required, and must be repeated. Late season application is required for some crops. Application rates vary considerably. On average, use 1-3 Tbsp. per gallon of water. Please call for crop-, disease- and pest-specific application information.

To lower soil pH value by 1 point, use 1-1.25 pounds per 50 gallons of water per 100 square feet.

The product discussed here can be obtained from The Green Spot by calling 603/942-8925. Here is the current pricing of this product for 1998...

Item no. PMSF1 ... 1 lb. shaker Micronized Sulfur (spray or dust) ... Can = $6.35
Item no. PMSF4 ... 4 lb. bag Micronized Sulfur (spray or dust) ... Bag = $12.00

EPA Flag-word: Caution. Product is toxic to fish.

Concentrated Liquid Sulfur

Concentrated Liquid Sulfur is an organic fungicide / miticide containing 52% elemental sulfur. It is labeled for use on potatoes, tomatoes and ten other fruit crops, including most citrus. It is effective in preventing several spots, rusts, certain rots, powdery mildew, scab and more.

Concentrated Liquid Sulfur is also labeled as a miticide. It is effective on rust mites and tomato russet mites. This product is not listed as being injurious to any crops.

In most instances very early application is required and must be repeated. Late season application is required for some crops. Application rates vary considerably. On average, use 1-7 Tbsp. per gallon of water. For tomato russet mite control use 0.5-1oz. per 340 square feet, depending on the size of plants. Please call for crop-, disease- and pest-specific application information.

The product discussed here can be obtained from The Green Spot by calling 603/942-8925. Here is the current pricing of this product for 1998...

Item no. PLSF16 ... 16 oz. Liq. Sulfur fung. conc. ... Btl = $8.85

EPA Flag-word: Caution. Product is toxic to fish.

Oil & Lime Sulphur Spray

Oil & Lime Sulphur Spray is a mixture of calcium polysulfides, 5% (a fungicidal sulfur), and horticultural oil, 80% (an insecticidal oil). It is labeled for use on six fruit crops, two nut crops, and a wide range of others including ornamental shrubs, deciduous trees and roses. It is effective in killing overwintering diseases such as spots, rusts, and blights, plus certain fungal spores. It also kills overwintering black, brown, Italian pear, rose and San Jose scale insects, mealybugs and spider mite eggs.

This product *will* kill most bio-control agents. However, since it is labeled for use on dormant plant materials and overwintering insects, plus, since it kills only on contact, the advantages will normally outweigh the disadvantages. Still, caution should be exercised as this product is considered dangerous.

Use 2.5ozs. per gallon of water. Call for more specific instructions and recommendations.

The product discussed here can be obtained from The Green Spot by calling 603/942-8925. Here is the current pricing of this product for 1998...

Item no. POLS16 ... 16 oz. (pt.) Oil & Lime Sulphur Spray ... Btl = $7.65

EPA Flag-word: Danger. Product is toxic to fish and mammals.

Lime Sulphur Spray

This product is the just like oil and lime sulphur, without the oil. Consequentially, it can be used *year 'round to provide the same control of the same diseases and most of the same pests. However, Lime Sulphur contains 30% calcium polysulfide versus the 5% in other. Much of this product's labeling is the same as for that above — other than the window of opportunity for its use. It should also be used with great caution.

*The Green Spot cannot in good conscience recommend the use of this product during any season but winter — despite what the label says.

Use 3 teaspoons to 1 pint per gallon of water, depending upon the time of year. For more information regarding this product's use, see Oil & Lime Sulphur Spray (above), or call us for more specific information.

The product discussed here can be obtained from The Green Spot by calling 603/942-8925. Here is the current pricing of this product for 1998...

Item no. PLSS32 ... 32 oz. (qt.) Lime Sulphur Spray ... Btl = $9.95

EPA Flag-word: Danger. Product is toxic to fish and mammals.

Rot-Stop

Rot-Stop is a product which helps correct the calcium deficiency which, in part, causes tomato blossom-end rot. It contains 9.2% Calcium derived from calcium chloride. Maximum chlorine level is 20.78%.

In 1995 we were asked if we had a product which would help cure this mysterious ailment we didn't at that time. All we could do is tell growers the culprit was either a calcium deficiency or irregular irrigation habits. In 1996 that all changed. We *were* able to help.

Use Rot-Stop on tomatoes, cucumbers, melons and peppers. Use at the rate of 2 oz. (4 tablespoons) per gallon of water. Apply during periods of rapid plant growth or after excessive rain or irrigation. Spray directly onto the foliage to the point of run off. The number of applications should be limited to avoid foliage burn. Repeat applications every 5-7 days, if needed.

The product discussed here can be obtained from The Green Spot by calling 603/942-8925. Here is the current pricing of this product for 1998...

Item no. PRS16 ... 16 oz. (pt.) Rot-Stop conc ... Btl = $5.95

EPA Flag-word: Caution.

Fire Blight Spray

Fire Blight Spray contains 21.2% Streptomycin sulfate. It is labeled for use on apples, chrysanthemums, dieffenbachia, pears, philodendron, pyracantha and roses.

It is effective in controlling fire blight, bacterial wilt, stem rot, leaf spot and crown galls.

Application rates and usage vary from crop to crop. Please call for specific information and recommendations. To mix: use 1 tablespoon per every 2.5 gallons of water.

The product discussed here can be obtained from The Green Spot by calling 603/942-8925. Here is the current pricing of this product for 1998...

Item no. PFBS2 ... 2 oz. Fire Blight Spray conc. ... Jar = $9.95

EPA Flag-word: Caution.

Phyton 27
Bactericide & Fungicide

Phyton 27® is a copper-based product which we believe will have few ill-effects on beneficials, is safe to plants (even after prolonged use), is highly systemic to provide lasting protection, leaves no unsightly residue and, according to the people we spoke with, Phyton 27 works incredibly well. This product can even be used on poinsettias after their bracts have changed color.

And, as far as the product's maker,

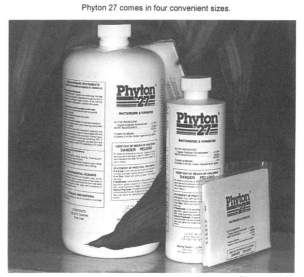

Phyton 27 comes in four convenient sizes.

Photo by M. Cherim

Source Technology Biologicals, Inc., they seem to really know and stand behind their product. By stand behind their product we mean that they put the toll-free number for their tech support service program on a Rolodex® card and provide it with every bottle. Fact sheets, bulletins, reports, etc., are also available. Source is the source for info about Phyton 27. And now we are too. Since we're dealing directly with the manufacturer, the support we can offer is seemingly limitless. Call us with your questions. We can send you any of the following literature absolutely free-of-charge: Product Data & Label Guide; the latest Phyton 27 News newsletter (which you can request a subscription to); the Poinsettia Technical Report; the Geranium Technical report; the Tropical Foliage Technical Report; the Rose Technical Report; one of the technical bulletins (Geraniums, Poinsettias, Calla Lilies, Orchids and Roses); or, perhaps one of the Use Guidelines Pamphlets entitled "For Control of... Fungal Diseases on Azalea & Rhododendron... Bacterial and Fungal Diseases on Orchids... Fire Blight and Apple Scab on Ornamental Apple, Pear and Mountain Ash... Bacterial and Fungal Disease on Begonias... Bacterial and Fungal Diseases on Tropical Foliage... Bacterial and Fungal Diseases on Poinsettias... Fungal and Bacterial Disease on Cyclamen... Fungal and Bacterial Diseases on Impatiens... Fungal and Bacterial Disease on Pansy, Petunia and Zinnia... Bacterial and Fungal Diseases on Geraniums... Fungal and Bacterial Diseases on Japanese Maple... Powdery Mildew and Botrytis on African Violets... Fungal Diseases on Roses For Cutting... Fungal Diseases on Bush Roses... Erwinia Soft Rot and Botrytis on Calla Lily," and two others, "For Bacterial and Fungal Control on Ornamental Crops (Azalea, Chrysanthemum, Exacum, Fuchsia, Gerbera, Holiday Cactus, Kalanchoe, Primula) and To Prevent Botrytis on Cut Flowers." If you have questions, we can answer them.

Phyton 27 contains 21.36% Copper-Sulfate Pentahydrate (copper as metallic 5.5%). It is considered a liquid copper complex comprised of tannates, picrates, ammonium formate and, as listed above, copper-sulfate Pentahydrate. It is labeled to suppress and control a huge assortment of fungal and bacterial diseases (see above). Call us for specific, plant by plant usage rates. There are way too many to list.

The product discussed here can be obtained from The Green Spot by calling 603/942-8925. Here is the current pricing of this product for 1998...

> **Item no. PP272 ... 2 oz. Phyton 27 conc. ... Btl = $14.95**
> **Item no. PP278 ... 8 oz. Phyton 27 conc. ... Btl = $42.85**
> **Item no. PP27L ... 33.7 oz. (liter) Phyton 27 conc. ... Btl = $99.75**
> **Item no. PP27G ... 128 oz. (gallon) Phyton 27 conc. ... Jug = $332.65**

EPA Flag-word: Danger. Product is toxic to fish and mammals. REI: 24 hours.

using tools

Hand held stone hammers? — been there, done that

Getting the product from the bottle to the plants may have been an enormous challenge for our cave-dwelling forefathers. But nowadays, thanks to modern technology and the advent of plastics, poisoning ourselves and our environment has become a very simplified process.

Okay, maybe we went a little overboard in our last paragraph in regard to "poisoning." We were correct, however, when we mentioned a "very simplified process." To disperse granular materials, liquids (even water), powders, dusts, etc. modern tools make it safer and more simple.

Check out the next section to get a brief idea of the types of application equipment available to today's growers. Certainly not everything is mentioned (awesome electrostatic sprayers, ATV spreader rigs, boom spray equipment, etc. are not), but the real common tools are.

You may notice that most of the tools we discuss and offer are for small-scale use. That's because a properly run pest and disease management program — even a very large-scale one — will make most broad-coverage treatments obsolete. Micro-manage and save time, materials and money. The nuclear bomb approach is going out of fashion.

Most application tools all have the same things in common: 1) they propel material from point A to point B while... 2) providing maximum coverage with... 3) as little waste as possible. They all make the job faster, easier and less costly.

For more detailed information and recommendations concerning the use of tools — primarily small scale application equipment — you are invited to call The Green Spot, Ltd. at 603/942-8925. 🧩

tools of the trade

A hose-end siphoning-type sprayer can be of immense value.

The Big Andy
Hose-End Sprayer

The Big Andy™ hose-end sprayer, unlike most, *removes* liquid from its reservoir without putting more back in. It does this by siphoning the contents out as the hose water passes over a small opening.

Needless to say, this hose-end sprayer is not practical for applying granular, water-soluble, or chemical fertilizers and such. It is, however, the easiest and most accurate way to apply nematodes in any area which can be reached by hose. This application is the main reason we discuss this lightweight, durable and economical product. Other applications would include the dilution and dispersal of other liquid-based products. Reservoir's contents, when full, dilutes to 20 gallons of ready-to-use product.

The product discussed here can be obtained from The Green Spot by calling 603/942-8925. Here is the current pricing of this product for 1998...

Item no. PHES ... 20 gal. Hose-end sprayer ... Each = $4.25

The RL Flo-Master
3 Pint Hand-Held Pressure Sprayer

For applying mist water to thwart spot infestations of spider mites, or to spot-spray with insecticidal soap, horticultural oil, fungicides, etc., this product is just what the doctor ordered.

The RL™ Flo-Master® 3 Pint Hand Pressure Sprayer is an excellent, high

All photos, above and both on facing page, by M. Cherim

quality, ruggedly constructed product designed to last for many years — possibly a lifetime if properly maintained.

Fill it with the liquid of choice, secure the screw-on top, pump it up about 20 times with the easy-to-use T-handle pump, then depress the thumb trigger to spray. It will continue spraying for several minutes until then air pressure is depleted. No more tired trigger finger. The adjustable nozzle allows the user to quickly select a from powerful stream of about 12' to an ultra fine mist, or anywhere in-between.

A small, pressurized sprayer can be conveniently carried by growers for the treatment of localized pest hot-spots.

In our opinion, this is the best small pump sprayer on the market, and the price is definitely right. [Actual working capacity - 2.5 pints (1.18 liters).]

The product discussed here can be obtained from The Green Spot by calling 603/942-8925. Here is the current pricing of this product for 1998...

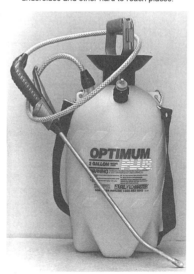

Wand sprayers are convenient for treating leaf undersides and other hard to reach places.

Item no. PHHPS ... 3 pt. RL Flo-Master sprayer ... Each = $10.95

The Optimum Plus
2 Gallon Pressurized Tank Sprayer

Here's another, quality, RL Corporation product. This ruggedly constructed tank sprayer is made to last.

The Optimum Plus® is a light-weight polyethylene sprayer features a patented elliptical design which makes it easy to handle, an easy-to-fill funnel top, a fluid level indicator, a 40" reinforced hose, a 14" brass wand, a carry strap for ease in portability, and an adjustable nozzle allowing the user to select from a powerful 18' stream to a delicate, ultra-fine mist.

Great for applying *Bt* to small to medium-size trees, plus spraying other products, insecticidal soap, horticultural oil, neem, fungicides, etc. [Actual working capacity - 1.5 gallons (5.7 liters).]

The product discussed here can be obtained from The Green Spot by calling 603/942-8925. Here is the current pricing of this product for 1998...

Item no. POPPTS ... 2 gal. Optimum Plus sprayer ... Each = $35.95

Hudson Suprema
4.5 Gallon Bak-Pak Sprayer

For larger spray jobs, the Hudson Suprema® 4.5 Gallon Bak-Pak® Sprayer is the choice of professionals.

This is a high-quality, commercial use backpack sprayer useful for applying *Bt*, horticultural oil, insecticidal soap, neem, fungicides or any water-consistency liquid.

Its features include: a comfortable trigger handle allowing adjustable flow; 1.75"x40" nonabsorbent, adjustable straps; a comfortable, kidney-shaped back; all-brass fittings and 18" wand; ultra-fine mist to 15' stream adjustable; a wide 4", screened fill opening; a noncorrosive, 4.5 gallon [actual working capacity] polyethylene tank, a strong, 36" chemical resistant Kem Oil® hose; and positive piston brass pump, offering a continuous 100 psi. It's built to last!

To use this sprayer: simply fill, close, pump and spray. No tools required. Empty weight 10.4 pounds.

The product discussed here can be obtained from The Green Spot by calling 603/942-8925. Here is the current pricing of this product for 1998...

Item no. PHSBPS ... 4.5 gal. Hudson Bak-Pak sprayer ... Each = $134.95

Dustin-Mizer
Dust Applicator

For applying Diatomaceous Earth, Micronized Sulfur Fungicide, Copper-Sulfate or any powdered substance, the Dustin-Mizer is absolutely the best product on the market; rated #1 by Organic Gardening®, and *should be* by the commercial

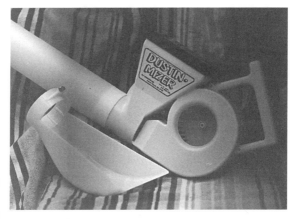

The Dustin-Mizer will blow you away — literally.

grower publications, as well (Greenhouse Grower®, GMPro®, NMPro®, GrowerTalks®, etc.). If your commercial operation is 20,000 sq. ft. or less, and you work with any dust product, you should seriously consider purchasing one of these handy tools. Greenhouse, nursery or field, it doesn't matter, 20k and under, and this product is a viable labor solution for thorough dust application.

It comes partially disassembled: with a crank body assembly, crank handle and 15.5" discharge tube. It will snap together in about 1 minute; and no tools are required. As an option, an extension/deflector attachment is available (shown in photo, above). This attachment allows the user better directional control. It is especially useful when under-leaf dust applications are desired.

The Dustin-Mizer is simple to use: fill the hopper (which unfortunately only holds about a pound of material but the tool is a *mizer*), snap the hopper cover on, aim and turn the crank. The combination of intermediate and impeller gears make for a blindingly-fast venturi action which in turn accelerates the screened dust at very high velocity through the discharge tube.

The Dustin-Mizer makes dust applications a breeze — literally. Actually, it's more like a hurricane-force wind (as you'll find out when you inevitably try it on yourself soon after the UPS® driver leaves). The Dustin-Mizer makes for economical, efficient, and even distribution. All parts are noncorrosive and built to last for many years. Rated number #1 by The Green Spot, too.

Photo by M. Cherim

The products discussed here can be obtained from The Green Spot by calling 603/942-8925. Here is the current pricing of these products for 1998...

> **Item no. PDMDA ... Dustin-Mizer dust applicator ... Each = $29.95**
> **Item no. PDMED ... Optional deflector extension ... Each = $7.50**

Red Devil Handi-Spred
Whirlybird Spreader

For applying *Nosema locustae*, *Bt-k* Dipel 10G, and other loose or granular products including seed, fertilizers, ice-melt material, etc., this product (see photo, inset top) is an excellent choice. But now it has an important new use concerning the application of certain bio-control agents. Certain products, *Neoseiulus =Amblyseius cucumeris* (see thrips controls) in loose bran, for example, can be evenly distributed over crops by sprinkling them out of the container (see photo, inset bottom). This is easily carried out on plants with broad surfaces or for small areas. For

Crank out the application procedure quickly with a spreader like that shown above or sprinkle product on plants the slow but accurate way as shown below.

Photo, top, by M. Cherim
Photo, bottom, by L. Gilkeson, courtesy of Applied Bio-nomics, Ltd.

larger areas, or on plants with small, strap-like leaves, a broad-cast spreader is really handy. (see Fig. 17, inset).

It comes fully assembled. All you have to do is fill the hopper (which holds from 3-6 pounds of material), select the distribution setting (4 range adjustable, complete instructions are included), turn the crank and go.

The Red Devil Handi-Spred efficiently, evenly and effectively covers up to a 12-14 foot wide swath per pass (less if you want), directly ahead of and two feet to the right, and to the left — at a 90 degree angle — of the operator.

This all plastic, noncorrosive, easy to maintain product is inexpensive, lightweight and built to last. It's a really smart buy.

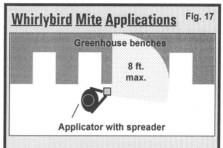

Whirlybird Mite Applications Fig. 17

Greenhouse benches

8 ft. max.

Applicator with spreader

Whirlybird-type spreaders should be seriously considered for the application of certain bio-control agents: Neoseiulus fallacis, Neoseiulus cucumeris and, perhaps, even Hypoaspis miles. They speed the delivery process while enhancing coverage accuracy. The spreader offered by The Green Spot has the capacity to spread up to a 12-14 foot swath. However, the applicator should be more gentle, turning the crank handle more slowly so as to minimize injury to the beneficials (see illustration). Experiment with coverages.

The product discussed here can be obtained from The Green Spot by calling 603/942-8925. Here is the current pricing of this product for 1998...

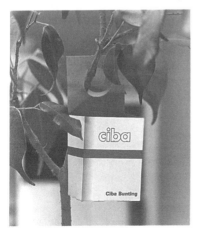

Boxes are great for tree applications.

Item no. PRDHS ... Red Devil whirly-bird spreader ... Each = $10.95

ciba Distribution Boxes

Here's another option for loose pupae and eggs or mites in flowable rice-hull, bran and vermiculite carriers. All of these things are sometimes difficult to apply in certain plantings: small, smooth leaves, hydroponics facilities, etc. These are all good examples of where these boxes would

Computer illustration, top, and photo, bottom, by M. Cherim

be of great value to many growers and interiorscapers. Another would be orchard mite applications.

Hang these distribution boxes from your plants' or trees' branches. Then add the material containing your beneficials. The bugs will crawl out of their own when ready.

Please note that these boxes should be purchased well ahead of time and painted green (outside only) if they're going to be used in public interiorscapes as red, white and blue tends to stand out and scream "look at me."

The product discussed here can be obtained from The Green Spot by calling 603/942-8925. Here is the current pricing of this product for 1998...

Item no. PDB10 ... 10 pk. dist. boxes ... Pack = $4.75
Item no. PDBC1 ... Case of 50 dist. boxes ... Case = $19.75

having to spray

We do what we have to, where we have to, when we have to

E very once in while a grower will call and confess that he or she feels the need to spray something with a little more kick than a biorational substance. And after talking to most of these people, we realize it's not what they really want to do, but what they feel is absolutely necessary. In fact, many times, the folks sound as if they're feeling a bit guilty or ashamed. But they shouldn't be. If they've been practicing the Green Methods, they've done an enormous amount of good already. Spraying a little bit of toxic material in a given location, with respect for their own bio-control program and the environment isn't all bad. We feel — coming from one business to another — that they may be doing what they feel is necessary to survive. In business, that means payroll and security for themselves and their employees. And there's nothing wrong with that.

We're not condoning the use of broad spectrum sprays used over a large, general area, but spot sprays; hitting only the pests when possible is very reasonable. If beneficials are on site, as they probably will be (even if not manually introduced), many, if not most should be spared.

There are only two substances that are going to be addressed in the next section. There are probably many other materials, in this book and not, which could fall into this category (such as hot pepper wax, see physical means). The main reason we only chose two items, though, is because of their residual period — or rather the amount of time they will affect the good bugs after the product is applied, which, in this case, is very short. Another reason is they target a multitude of pests and can be used in many situations.

Other products which could have been placed in this section, such as pyrethrins, last too long after they're sprayed. Pyrethrins, for example, remains dangerous to the good bugs mentioned in this book for about seven days.

For more detailed information and recommendations concerning the use of items you might need to spray in a pinch, you are invited to call The Green Spot, Ltd. at 603/942-8925. ✄

last resorts

Safer Insecticidal Soap

Safer® Insecticidal Soap is made from the potassium salts of fatty acids found in certain plant oils.

It works on contact by breaking down the target pest's cuticle — promoting dehydration and, ultimately, death.

Safer Insecticidal Soap, used indoors or out, is effective on, aphids, cabbageworms, earwigs, flea beetles, lace bugs, leafhoppers, mealybugs, psyllids, sawfly larvae, scale crawlers, squash bugs, thrips, spider mites, whiteflies and more.

Dish soap may work, but it is not the same thing, and it is not labeled for such use.

It is labeled for use on nearly all crops: bedding plants, flowers, foliage plants, fruits and nuts, herbs and spices, ornamentals, pot plants, trees and shrubs, vegetables and others.

Safer Insecticidal Soap breaks down fast, usually within 48 hours [okay for beneficials when dry (2 hrs.)]. If applied carefully, it may be effectively used around most bio-control agents. Please note, however, being a broad spectrum, contact insecticide, this product *will* harm most beneficials if it is applied directly on them. Therefore, judicious, small-scale spot applications are recommended in situations where bio-control agents are already on site.

Insecticidal soap is highly recommended for use a few days prior to predator

Photo by M. Cherim

and parasitoid inoculations, especially if pest numbers are high. In this capacity, because of its fast breakdown time, insecticidal soap will bring pest numbers down to more manageable levels, thus increasing the speed and odds of your beneficials' effectiveness.

Please note: Insecticidal soap is phytotoxic to certain plants. Therefore, test plots should treated prior to broad usage.

Safer Insecticidal Soap, in the concentrated form we carry, is quite economical, fast working, safe and effective. To use: Mix approximately 2.5ozs. per gallon of water. Repeat applications weekly, 2-3 times, or as needed.

The product discussed here can be obtained from The Green Spot by calling 603/942-8925. Here is the current pricing of this product for 1998...

Item no. PSISC16 ... 16 oz. (pt.) Safer insecticidal soap conc. ... Btl = $10.25
Item no. PSISC32 ... 32 oz. (qt.) Safer insecticidal soap conc. ... Btl = $17.95

EPA Flag-word: Caution. Product is hazardous to beneficials. REI: 4 hours.

Horticultural Spray Oil
Dormant & Summer Superior Oil Formula

This Bonide® product contains a self-emulsifying, highly paraffinic petroleum oil [98%, U.R. min. 92% superior type (contains Sunspray®)] used specifically to coat and smother certain pests of deciduous fruit and shade trees, roses and ornamental shrubs. Coating pests with horticultural oil blocks the passage of air through their spiracles (breathing holes), thus killing them.

It is labeled for use against overwintering eggs of European red spider mites, scale insects, apple aphids (not rosy aphids), bud moths, leafrollers, red bugs, codling moth larvae, pear psylla (adults), blister mites, galls, whitefly nymphs, and mealybugs.

This product may used at four different times: during the *dormant stage* (buds not showing green), the *green tip stage* (when bursting blossom buds show 1/8"-1/4" green), the *delayed dormant stage* (when leaves of blossom buds are out 1/4"-1/2"), and this product can also be used during the *growing season* as a ultra-fine summer oil treatment.

Use 2.5-8ozs. of this concentrated oil per gallon of water, depending upon the time of use, pest, and tree or shrub to be treated. Call for more details, precautions, etc. Please note: horticultural oil *will* prove harmful to benefi-

Oil is an extremely effective product, but it can be messy.

cials if it contacts them directly. However, when used as a dormant oil, this product will impact very few. [Okay for beneficials when dry (8 hrs.).]

A special note to interior-scapers using this product: using a drop-cloth, cover the floor or carpet at your account. A carpet can become terribly stained or, worse, someone could slip on an oily floor. Another idea, for anyone, not just interiorscapers, is to apply horticultural oil with a brush — brushing it directly onto scale insects, etc. This will offer optimum control of the product; no mess, no overkill, no waste, plus it will provide better contact and, thus, better control.

The product discussed here can be obtained from The Green Spot by calling 603/942-8925. Here is the current pricing of this product for 1998...

Item no. PHSOC16 ... 16 oz. (pt.) Horticultural spray oil ... Btl = $7.95
Item no. PHSOC32 ... 32 oz. (qt.) Horticultural spray oil ... Btl = $10.95

EPA Flag-word: Caution. Product is toxic to fish / hazardous to beneficials. REI: 4 hours.

other practices

The consequences of all actions must be considered

Using the Green Methods in the interiorscape, greenhouse, farm, etc. is great, but what we do outside that particular arena may play a part in the success of the overall picture.

The section which follows details some products which may provide solutions to the periphery by not inhibiting the effectiveness of the other organisms and/or products mentioned.

One example: the fly traps. If you have a total farm operation, say a greenhouse range, field crops and livestock, you may have flies. Attacking those flies with poisons may have a negative impact on the other, more natural farm operations.

Another example: the honeydew substitute, Biodiet. Perhaps you're thinking about getting ladybugs and want to make the most of your purchase, but you don't have a terrific aphid population. This product may encourage more of the ladybugs to stick around for a while longer.

For more detailed information and recommendations concerning other practices, you are invited to call The Green Spot, Ltd. at 603/942-8925. ❊

other goods

Rescue!
Disposable Fly Control Trap

These Rescue!™ Disposable Fly Control Traps are really simple to use: cut the seal, pull out the top, crush the ampule, add water and hang. That's all there is to it.

Each trap will capture anywhere from 15,000 to 20,000 flies. No mess, no bother. And they *will* catch flies effectively for months, depending on the severity of the fly problem in your area. When it's full, close the top and throw it away.

Their applications range from the backyard to the horse barn, from the kennel to the compost pile. Feedback has been very positive. And it has been tried and proven, again and again, in all of the above-named situations, with consistent results.

Effective, efficient, easy to handle, and easy on the budget, too. Start out by trying 1 trap for every 2-4000 sq. ft.

The product discussed here can be obtained from The Green Spot by calling 603/942-8925. Here is the current pricing of this product for 1998...

Item no. PDFT ... Rescue disp. fly trap ... Each = $5.25

Solar-Type
Reusable Fly Control Trap

For large-scale commercial use, the corrosion resistant, aluminum, Solar-Type Reusable Fly Control Trap works best. It's large, sturdy, and will last for years.

Artwork courtesy of manufacturer

Solar-type fly traps work by luring flies to a feeding bowl, from which they fly upward through a metal cone into the trap. Once inside, they can't get out. Trapped, the flies eventually die of dehydration as they are "cooked" in the sun.

Each trap comes with a 4-5 week supply of a specially formulated yeast based "stink bait" which must be mixed and fermented prior to use (a 1 gallon milk jug works well for this process). Replacement bait packs are available (see below). Hint: Save freight money by buying enough bait for the season or, you may wish to try concocting your own formula.

Other baits such as dog food (in summer) or rotted fruit (in spring and fall) may attract and trap yellowjacket wasps. Deer flies and horse flies may also be trapped in solar-type traps. However, *thorough* control of the above-mentioned pests is not likely. (We discuss a special yellowjacket trap in this section, see below.)

In a large fly control program, the best results will be obtained not only by using fly parasitoids (see page 144) to reduce on-site fly numbers, but by trapping inbound flies, too. We've gotten a lot of feedback stating, matter-of-factly, that the integrated approach to fly control works; with an average cost: about $10 per horse or cow for the entire summer, but will depend upon farm size. Use one or two of these traps per every 100 horses or 50 cows.

The products discussed here can be obtained from The Green Spot by calling 603/942-8925. Here is the current pricing of these products for 1998...

> **Item no. PSTFT ... Solar-type fly trap ... Each = $59.95**
> **Item no. PSTFTB ... Solar-type fly trap bait ... Bag = $7.95**

Rescue! Disposable Yellowjacket Control Trap

Yellowjacket wasps, a beneficial social species useful as scavengers and predators of pestiferous insects, can be a real annoyance to people. They are greatly feared by many, and can sometimes be a very real threat. That's

Photo courtesy of Arizona Biological Control Company, computer scan by M. Cherim

where the Rescue!™ Disposable Yellowjacket Control Trap comes in handy. Using a special attractant, this trap draws yellowjackets inside. There, they are trapped. No mess. No fuss. No chemicals. No painful stings.

Other ways to control yellowjackets without chemicals or trapping includes: spraying them with cold water from a hose (which doesn't kill them, but *does* knock them senseless for quite a while); or, in rural areas by placing honey next to their in-ground nest in the evening so that animals such as raccoons and skunks, drawn by the honey, may dig up and destroy the nest for more. Regardless of what method of control you try, we urge you to be careful. Yellowjacket wasps *can* deliver a painful sting if they feel threatened. If you are hypersensitive to stings, or suspect that you may be, please consult a professional. Call for more information.

The product discussed here can be obtained from The Green Spot by calling 603/942-8925. Here is the current pricing of this product for 1998...

Item no. PDYJT ... Rescue disp. yellowjacket trap ... Each = $5.25

Rescue! Soldier Bug Attractor

The spined soldier bug (see illustration, below), *Podisus maculiventris*, is an outstanding, North American native predatory true bug. *P. maculiventris*, with its needlelike proboscis, preys on various stages of over 35 garden pests including: Colorado potato beetles, corn earworms, Mexican bean beetles,

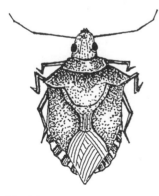

tomato fruit worms, cucumber beetles, armyworms, cabbage loopers, imported cabbage worms, lace bugs, flea beetles, webworms, tarnished plant bugs and others. It also preys on various stages of over 18 tree and foliage pests including: Gypsy moths, tent caterpillars, codling moths, leafrollers and others. We're very impressed with this predator. Moreover, we're extremely excited that we may soon have an opportunity to offer these predators (as actual nymphs and/or adults) for sale;

Artwork, top, courtesy of manufacturer
Illustration, bottom, by M. Cherim

sometime in the near future anyway... we hope.

The Rescue!™ Soldier Bug Attractor, which can be planted in the ground or hung in a tree, attracts *P. maculiventris* by means of a powerful phero-mone or sex lure. The lure draws the female soldier bugs, who come-hither to mate. When they get to the lure all they find is this big, yellow, plastic thing. Frustrated, if pests are abundant, they will eat away their blues. Put out these lures as early as possible in the spring (1st to 3rd week of April). Note: *P. maculiventris* adult emergence coincides with bud-burst of red and silver maple trees, yellowing of pussy willow catkins, or pink bud of saucer magnolia. New generation nymph arrival coincides with bud-burst of Washington and cockspur hawthorn, and fringetree. Put lures out at these times for best results. Use 1 lure per every 200 square feet of growing area, or 1 lure per every orchard tree. Large trees, use 2 or more lures. Best results will be obtained if used outdoors. The pheromone attrac-tant remains active for about 60 days. Replace lures as needed.

This product is recyclable. However, if your local recycling center will not accept this product, the manufacturer, Sterling International, Inc., will accept the product back (details are included in the box).

The product discussed here can be obtained from The Green Spot by calling 603/942-8925. Here is the current pricing of this product for 1998...

Item no. PSBA ... Rescue soldier bug attractor ... Each = $5.45

Sure-Fire Ladybug Lures

In addition to using Biodiet™ and Bioblend (both listed in this section), ladybugs can be drawn to your property by using these innovative lures. Sure-Fire™ Ladybug Lures contain a special kairomone scent — an allelochemic substance which sends a clear message to ladybugs in sur-rounding areas that aphids [foodstuff] are available.

Whether or not aphids are really available for the ladybugs is irrelevant, for they will come anyway.

Artwork courtesy of respective manufacturers

And they may nip away a newly arrived aphids or other pests which may happen to be on site.

They come 3 to a box. Use 1 lure per every 200 square feet of growing area early in the season. Lures remain active for about 2-3 weeks. Best results will be obtained if used outdoors.

The product discussed here can be obtained from The Green Spot by calling 603/942-8925. Here is the current pricing of this product for 1998...

Item no. PLBL ... 3 pk. Ladybug lures ... Pack = $4.95

Biodiet
Beneficial Insect Food

Biodiet™ is an all-natural, scientific formulation containing *Kluyveromyces fragilis* yeast, a milk-whey substrate (hydrolysate of casein protein), sugar, and 1.5% cornstarch as an anti-caking agent. This honeydew substitute formula (or a similar product) is commonly used in commercial insectaries to aid in the mass propagation of green lacewings: *Chrysoperla* spp. This product's special properties are known to increase the commercial lacewing egg-laying production significantly. Using this mixture may enhance the productivity of several species of natural *and* purchased beneficial insects. It can help keep them in your garden, in your greenhouse, or on your farm for a longer period of time. There, they can prey on your pests. It also aids in *attracting* several beneficial insect species. Biodiet works by partially satisfying the beneficial insect's protein/carbohydrate requirements:

Biodiet is appreciated by many beetles — especially Crypts.

Photo by M. Cherim

allowing lacewings to lay more eggs, and to prompt ladybug beetle species to stay in place and consume more pests — which must be present for *them* to lay eggs. If you purchase green lacewing adults or any of the various beetles shown in this manual, you may want to add Biodiet to your shopping list.

To use: In a sealable container, combine 1 part Biodiet with 5 parts hot water. Mix thoroughly, then allow the mixture to cool. Soak up the formula with wads of paper towel, pieces of sponge or a similarly absorbent material. Warning: Do not use cotton balls, as they may contain pesticide residues. Strategically place the pieces in your target area: 1 every 100 sq. ft. When the pieces dry out, re-hydrate them — which can be done until the smell and coloration of the formula is no longer discernible; at which point, repeat the entire process, from the beginning.

Another way to prepare Biodiet is to make a paste (as is supplied with many of our little critters). To do this mix roughly 10 parts of Biodiet with 1 part cool water. Stir until you have a well-mixed paste. Smear this paste on little squares of cardboard, etc. and place them throughout the area.

Tip: The absorbent pieces may develop a mold, in which case, repeat the process. The mold develops from water-borne microorganisms growing on the Biodiet, much like an agar. If mold develops frequently, using distilled water may help.

Tip: Prior to releasing purchased ladybugs or adult lacewings, place several "dinner sponges" (or smears of paste) at the release site. It will help keep the insects healthy, more prone to egg-laying, and in the area longer.

Tip: A little bit goes a long way; mix only small portions at a time. Store unused portions of pre-mixed formula in a sealed container in your freezer: this will help maintain its freshness. This one, 8oz. bag should last at *least* a complete growing season for a 1000 square foot area.

Please note: Biodiet does not interfere or inhibit the function of your bio-control agents. However, it may draw ants. If ants could be a problem, use this product sparingly.

The product discussed here can be obtained from The Green Spot by calling 603/942-8925. Here is the current pricing of this product for 1998...

Item no. PBD8 ... 8 oz. bag of Biodiet ... Bag = $4.50

Bioblend
Beneficial Seed Mixture

Sow this special mixture of seeds in or around your cash crop and treat your bio-control agents to something special.

The hoverfly shown in this photo is a valuable natural predator of aphids.

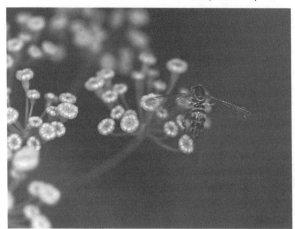

The flowering plants in Bioblend provide nectar and pollen for your beneficials, assuring they have a balanced diet, a place to seek refuge and reproduce, and a place to feed and feel comfortable.

Bioblend, in this regard, serves as a trap-crop. It lures pestiferous species away from your valuable crops. If a battle between good insects and bad is going to take place, it is a good idea to offer them a battlefield in which to war — a battlefield *away* from your plants.

Bioblend may also attract indigenous beneficials such as the syrphid or hoverfly shown in the (see photo, above).

Bioblend contains a mixture of alfalfa, alyssum, carrot, celery, clover, daikon, dill (see photo, above), eriogonum, fasiculatum, fennel, gypsophila, nasturtium, radish and yarrow. (Mixture may vary.)

Use along isles and under benches in a greenhouse, outside the greenhouse (but not near ventilation intakes), also surround fields, border gardens, or create a haven or special "diversity area" with its own microenvironment.

Sow at the rate of 1oz. per 100-200 square feet. We recommend planting enough Bioblend to equal 2-5% of your cash crop area.

The product discussed here can be obtained from The Green Spot by calling 603/942-8925. Here is the current pricing of this product for 1998...

Item no. PBBSM ... 1 oz. of Bioblend ... Oz. = $.95

White Cedar Bat House

Bats consume approximately 300-500 mosquitoes in one evening (in addition to myriad other insects such as moths and beetles). In three months time that equates, on average, to 36,400 mosquitoes. Multiply that by the 18 bats which possibly be housed by one white cedar bat house and you will clear your property of approximately 655,200 mosquitoes in three months.

Bats have a bad rap; they're not half the blood suckers that mosquitoes are.

655,200 mosquitoes barely puts a dent in the mosquito population is many areas. However, it helps. If used in combination with Mosquito Dunks® (see page 188), you may find that you'll actually be able to venture out of the house on a summer evening without whirling around in circles, puffing maniacally on a cigar.

Bats are our friends. They are feared by many people. However, they are creatures which are useful to humankind. Unless you are a mosquito lover, be neighborly, give your bats a home in the country with one of these natural, white cedar bat houses.

Use 2-4 bat houses per acre, or more in heavy mosquito country. Suspend each bat house on the side of a tree or structure, facing east, approximately 12-15' off the ground in an area with a source of water. It can take 1-3 years for bats to establish themselves in a new bat house Complete instructions and hanging hardware are included with each unit.

The product discussed here can be obtained from The Green Spot by calling 603/942-8925. Here is the current pricing of this product for 1998...

Item no. PSCBH ... White cedar bat house ... Each = $7.95

Both photos, above and facing page, by M. Cherim

Drax Ant Baits

Drax® Ant Baits come in two special formulas: a sweet mint-Gel with boric acid (used to kill sugar-eating ants); and a special Protein Formula bait with boric acid (used to kill grease, fat and meat-eating ants). Determine what ants you have by setting out a snack of jelly or peanut butter prior to ordering. If your ants go for the jelly, they are sugar-eating ants; if they go for the peanut butter, they are protein-eating ants.

Using Drax Ant Baits is an effective way to provide long-term control of ants indoors and out.

Both formulas, containing Orthoboric acid as the active ingredient, are consumed by foraging ants, brought back to the colony, fed to the queen and to the young. The ants soon die. Moreover, the dead ants will often be consumed by others in the colony, and they, too, will perish.

Drax Ant Baits are long-lasting, effective, and inexpensive. Call for more information.

The products discussed here can be obtained from The Green Spot by calling 603/942-8925. Here is the current pricing of these products for 1998...

Item no. PDABG12 ... 12 oz. bottle Drax Ant-Kil Gel ... Btl = $4.75
Item no. PDABPF4 ... 4 oz. tube Drax Ant-Kil PF ... Tube = $4.95

EPA Flag-word: Caution.

These products have been discontinued by the manufacturer. Available while supplies last. The replacement product(s) has not yet been determined, but may end up being a duel-tube syringe containing both formulations.

The Business End

**because it all
comes down
to money**

why we do it

We've been honest with you so far — so we'll continue

We've been asked why we do such a "bulky" publication and how we afford it. When the bill comes in, we ask ourselves the very same thing. And the answer, simply put, is that we can't afford not to.

When we first got into this business (a lateral move from being a grower), information about bio-control and IPM was hard to come by. Price sheets flowed in like the Bay of Fundy's tide, but getting rock-solid details on the how-tos and what-ifs was like pulling teeth.

Moving into this business we saw exactly what the industry needed: facts. How else were we going to get the right word out. We didn't want to mislead our customers or give them conflicting and flaky information. Who needs that? But we didn't even have a book adequate for our own use and needs. How could we possibly retain the information for our customers when they called with questions.

Opportunity kicked the door down — good thing we were home and wide awake!

But we are left with one problem: no money — as writing, publishing and book sales are not a profit center for The Green Spot. Our operations money comes from the sales of the products we discuss in this book.

If you purchased this book, we broke even. That doesn't include payroll, advertising, simple business expenses (i.e., telephone, rent, etc.). If we relied on book sales only, the price of this publication would have to be in the neighborhood of $35.00, and we have to come up with an additional eleven titles every year.

If you enjoy this book, we're glad. If you're learning things, that's terrific. If you'd like to see us continue with our efforts — because the truth costs money — we need your help. If you feel you can use something in this book and need to make a purchase, contact us. It's what we do for a living — not publishing.

If you want to talk, need help or advice, call us at 603/942-8925. We won't sell you hard; honesty is the premise of this business. If you want to order supplies, need support, or have questions about the business end of The Green Spot, please read the section which follows. Oh, and thank you very much.

business details

How to Order

Ordering from The Green Spot is easy. The preferable method is by telephone. We can both communicate nearly simultaneously that way. If you have questions, we provide almost instantaneous answers. We will make recommendations geared towards getting the job done in the fastest, most economical manner possible — even if it involves a recommendation which does not mean a sale of product at that time (which hopefully will flabbergast you, as the customer, so much by our lack of greed that you'll ultimately be a long-standing regular).

If you know exactly what you need (because we spelled out the necessary information so well in this book), you may choose an alternate ordering method such as by fax, e-mail or 24 hour voice message. If you need a confirmation of your alternately placed order, let us know when ordering.

Please note that we do not provide consultation by any means other than by telephone during regular business hours, unless, however, you are willing to pay for it. *Written* consultation is billed by the page at $19.00 per. But this is not a preferred method.

To get in touch with us, please note the following numbers and business hours (also note that we have strict deadlines for order live materials, see next):

TELEPHONE: 603/942-8925 - Available M-TH 8 AM - 5 PM, FRI 8 AM - 4 PM est.
FAX: 603/942-8932 - Available 24 hours/day, 7 days/week.
VOICE MAIL: 603/942-5027 - Available 24 hours/day, 7 days/week.
E-MAIL: GrnSpt@internetMCI.com - Checked TH and FRI only.

Ordering Deadlines

Dealing with specialized products such as bio-control agents, special considerations must be made to ensure product integrity. These enforced deadlines — which apply for all order cancellations as well — are poised with your interests in mind. Please conform to these deadlines to avoid problems and delays (see next page):

BIO-CONTROLS AGENTS ... FRI at 3:00 PM est. for delivery the next shipping week.

EXCEPTIONS INCLUDE ... Aphidius colemani, Eretmocerus eremicus, leafminer parasitoids and bumblebees. For these items the deadline is TH at 3:00 PM for delivery the next shipping week. Orius insidiosus should also be ordered by this time to be safe, but may be waived during the spring and summer months.

ALL OTHER PRODUCTS ... If ordered by 4:00 PM est. will ship the same day, except of Fridays.

Standing Orders

To avoid having to remember deadlines and such, we welcome standing orders. Standing orders will allow us to ship product(s) to you at intervals you will set up ahead of time. The orders will be numbered on the box and on the receipt or invoice (1/2, 2/2, etc.). This is a great value-added service to today's busy growers. And there is no additional charge for this service. We have an extremely reliable system of performing this service — one less detail for you, the busy grower or interiorscaper, to deal with. Moreover, billing is the same as it usually is, one bill for each shipment. There is never a need to pay for a whole series of standing orders ahead of time. In other words you don't have to break the bank in the name of convenience and peace of mind. Please note, we are typically reluctant to set up standing orders for a period of over 13 weeks, but that may depend on your individual circumstances. Please call us for details.

Paying for your Order

We have a very unusual policy here at The Green Spot: we will give *all of our customers, old and new alike, an automatically applied credit line of $300.00 (so, yes, we will accept your purchase order).

This policy is one our customers have enjoyed for the past four years. And we'd like to say thanks to the *vast* majority of our customers for not taking unnecessary advantage of our trust. It is because of your decency that we are able to continue offering this extension of our trust. We will still accept Visa®, MasterCard® and Novus℠ (Discover®) cards as a method of payment. But if you want to be billed, that's okay too. Here are our terms:

RETAIL / INDIVIDUAL CUSTOMERS: Net 15 Days

COMMERCIAL / INDUSTRY BUSINESS AND .EDU CUSTOMERS: Net 25 Days

DEALER / WHOLESALE CUSTOMERS: Net 30 Days

*Failure to make timely payments will negate this policy. Applies to U.S. customers only.

Discounts

There are three ways to obtain discounts from The Green Spot: our V.I.P. program; our .Edu discount program and; our normal volume discounts offered in the published pricing in this manual. Below we will explain the first two mentioned...

THE V.I.P. *PROGRAM, TIER I: Spend $1,000.00, including freight costs, call us for confirmation and, if confirmed, you'll receive a 5% discount on all line items, not shipping, handling, etc., for the remainder of the calendar year. You must keep track of your spending as we will not automatically apply this discount.

THE V.I.P. *PROGRAM, TIER II: Spend $10,000.00, including freight costs, call us for confirmation and, if confirmed, you'll receive a 10% discount on all line items, not shipping, handling, etc., for the remainder of the calendar year. You must keep track of your spending as we will not automatically apply this discount.

THE .EDU DISCOUNT *PROGRAM: Qualifying schools and universities may receive a 5% cooperation discount for the exchange of educational information pertaining to certain research, trials and studies. A contract must be signed before the discount is approved. For a contract copy, please give us a call at 603/942-8925.

*To qualify for any discount program, payments must be timely and accounts current. Discounts may be used for 1998 pricing only. Orders shipped after December 31st, 1998 will not qualify. Please call for details.

Ladybugs

With exception to the 100 and 500 count jars of ladybugs, all *Hippodamia convergens* orders will be shipped separately from other orders. (Billing may be separate too.) Shipments will be made directly from the collectors in California against our shipment and release permits. This is a new policy instituted to ensure a whole new level and consistency of freshness and quality. The associated costs with these separate shipments will be minimal and, in many cases, less than if shipment from our location because of the alternate shipping methods used.

Shipping Methods  United Parcel Service

We aim to provide the most reliable shipping methods available. Based on our experiences, the methods which follow meet, and often exceed, our expectations. We realize that some of these methods can be costly, but after utilizing "budget" couriers, and having problems with more than 20% of our shipments, we've come to the conclusion that, as the saying goes, you get what you pay for.

NONPERISHABLES: standard deliveries will be made by UPS® via their GroundTrac℠ service for contiguous U.S. delivery points. This service can take from 1 to 8 business days for delivery. If you're in a hurry, UPS offers 2 and 3 day guaranteed delivery options at a slightly higher price.

PERISHABLES: standard deliveries will be made by UPS® via their Next Day Air℠ service for all U.S. delivery points. This service is guaranteed for delivery the next business day. Perishable shipments will not be sent on Fridays. If your receiving location is within New England (CT, MA, NH, RI, VT and most of ME), you may choose to have your shipment sent via the nonperishable method described above [UPS GroundTrac]. If delivery is not made on the next business day, though, our product guarantee will be cancelled. This is your choice — it can save you some serious money, if you're willing to take the risk. Based on our experiences, the odds are greatly in your favor, especially during good weather when deliveries are made to business locations. Ladybugs, other than the 100 and 500 count units, will typically be sent via USPS Priority Mail.

DELIVERY RATES: will reflect the actual charges published by the courier used. They are not inclusive of our extra charges as described below. Guarantees will not be honored for late shipments of perishables unless the courier accepts the charges. We cannot force our policies on the shipper, and we therefore cannot accept responsibility for strikes, acts of God, mechanical problems, etc. To put your mind at ease, though, UPS has proven itself an extremely reliable courier.

Extra Charges

The enormous cost of packaging materials and labor dictates that we must levy extra charges to our standard shipping costs, as follows:

PERISHABLES REQUIRING SPECIAL PACKAGING: $3.50 per shipment will be added. This is waived for orders of $100.00 or more, and for shipments made on our behalf, unless we are charged by the sender.

NONPERISHABLES REQUIRING CORRUGATED BOXES: $1.50 per shipment will be added. This is waived for orders of $100.00 or more, and for shipments made on our behalf, unless we are charged by the sender.

SMALL ORDERS: of less than $15.00, including credit card, prepaid and billed orders will be assessed $1.50 per applicable order.

OTHER CHARGES: for C.O.D.s, Saturday deliveries, address corrections, etc. will be added or billed at a later date when applicable.

Our Guarantees

BIOLOGICAL PEST CONTROL AGENTS: All living materials, shipped by overnight *method, are guaranteed to arrive alive and healthy. Though all organisms are thoroughly inspected prior to being handled by the courier, and packaged, handled and shipped in a manner consistent with their requirements, an occasional loss is expected. In such a case, if a claim is filed within 72 hours of delivery, we will: 1) replace the affected portion** or entire order at no charge; or 2) offer a refund for the value of the affected portion** or entire order, if desired, on prepaid orders; or 3) credit your account (or credit card account) for the value of the affected portion** or entire order. In some cases, we will ask you to return the organisms, at our expense, or voluntarily at your expense, for investigative research to determine why the organisms perished or became unhealthy, as this is of utmost concern. Do NOT, under any circumstances, discard expired organisms prior to contacting our office for instructions. *Overnight shipping methods consist of both guaranteed and non-guaranteed services, as defined by the courier company. However, our live and healthy delivery guarantee applies to non-guaranteed shipping methods only when those methods result in overnight delivery. **A loss of some specimens in transit is considered normal. Losses of 15% (20% for lady beetles, true bugs and adult parasitoids), or less, will not be compensated. However, units are usually overpacked to compensate for normal losses incurred in transit.

ALL OTHER PRODUCTS: All nonliving products are guaranteed against breakage, missing parts, or other manufacturing defects. In the event of a claim, after the entire product, or unused portion, is returned, we will: replace the product; or issue a refund up to 100% of the product value, if desired, on prepaid orders; or credit your account (or credit card account) for up to 100% of the product value. Merchandise must be returned, and return authorized, prior to any compensation. Claims must be made within 2 weeks of product delivery. Individual manufacturers may have warranties and the such which supersede our guarantee.

Our Disclaimer

Due to the profound number of contributing factors involved with biological pest control & IPM, we will not guarantee, expressed or implied, the efficacy of any organism, product or service offered in our catalog, in any mailing, over the telephone or in any other manner. Efficacy is dependent upon the experiences of trial and error of the end-user. Recommendations, suggestions and/or advisories made in our catalog, in any mailing, over the telephone or in any other manner, are only recommendations, suggestions and/or advisories based on past or present experiences with our own research and development program(s) or, based on the successful experiences of other end-users with circumstances similar to those described by you, your pest control specialist, purchasing agent and/or other designated representative. Any and all product claims made in any way are based upon documented or undocumented field trials, past or present experiences and/or, on the labels of manufactured items. Any and all products, including live organisms, offered for resale, have been tested by others in agriculture, by insectary entomologists, by product manufacturers, by private testing facilities and/or, by any combination thereof and approved on state and federal levels, by either the EPA or USDA/APHIS/PPQ/BATS. The Green Spot, Ltd., its affiliates, dealers, resellers and/or assigns will not be held liable for any damages, direct or consequential, from the use or misuse of any product or organism made available in any way, extending beyond the purchase price. The buyer, by act of accepting any and all goods, agrees to assume any and all risks. You, however, may have certain rights governed by the laws in your state or country which offer protection above and beyond the scope of our protocols. Contact your consumer affairs department for more information.

Database

If, for any reason, you don't want to be in our computer database, please contact us so that we can remove your name. As company policy dictates, we will not release your name, address, phone number, etc., to anyone without your consent — ever. Our database is *strictly* confidential. We respect the folks who've given us their private information. We wish those large mail order companies would offer us the same courtesy.

Special Notice for 1998

The Green Spot, Ltd. will be closed, in addition to weekends and major holidays, July 2nd and 3rd, 1998.

The reason may sound a bit weak, but all of The Green Spot's employees want to take some or part of the week before July 4th off. As respect for our employees is of extreme importance to us, and we don't want to have any sacrificial lambs amongst the flock, we will simply close our doors during these two days when we would normally be open.

In advance, we want to apologize for any inconvenience this may cause you. Please place your orders before this time if you want a regular shipment the week following. We will also ship the week before, as usual.

Thank you, in advance, for your understanding.